MODELING, CONTROL, AND OPTIMIZATION OF NATURAL GAS PROCESSING PLANTS

MODELING, CONTROL, AND OPTIMIZATION OF NATURAL GAS PROCESSING PLANTS

WILLIAM A. POE

Senior Principal Technical Consultant,
Schneider Electric Software, USA

SAEID MOKHATAB

Gas Processing Consultant, Canada

ELSEVIER

AMSTERDAM • BOSTON • HEIDELBERG • LONDON
NEW YORK • OXFORD • PARIS • SAN DIEGO
SAN FRANCISCO • SINGAPORE • SYDNEY • TOKYO

Gulf Professional Publishing is an imprint of Elsevier

Gulf Professional Publishing is an imprint of Elsevier
50 Hampshire Street, 5th Floor, Cambridge, MA 02139, United States
The Boulevard, Langford Lane, Kidlington, Oxford, OX5 1GB, United Kingdom

Notices
Knowledge and best practice in this field are constantly changing. As new research and experience broaden our understanding, changes in research methods, professional practices, or medical treatment may become necessary.

Practitioners and researchers must always rely on their own experience and knowledge in evaluating and using any information, methods, compounds, or experiments described herein. In using such information or methods they should be mindful of their own safety and the safety of others, including parties for whom they have a professional responsibility.

To the fullest extent of the law, neither the Publisher nor the authors, contributors, or editors, assume any liability for any injury and/or damage to persons or property as a matter of products liability, negligence or otherwise, or from any use or operation of any methods, products, instructions, or ideas contained in the material herein.

Library of Congress Cataloging-in-Publication Data
A catalog record for this book is available from the Library of Congress

British Library Cataloguing-in-Publication Data
A catalogue record for this book is available from the British Library

ISBN: 978-0-12-802961-9

For information on all Gulf Professional Publishing
visit our website at https://www.elsevier.com/

Working together
to grow libraries in
developing countries

www.elsevier.com • www.bookaid.org

Publisher: Joe Hayton
Senior Acquisition Editor: Katie Hammon
Senior Editorial Project Manager: Kattie Washington
Production Project Manager: Kiruthika Govindaraju
Designer: Maria Inês Cruz

Typeset by TNQ Books and Journals

CONTENTS

ACKNOWLEDGMENT

This book would not be possible without the support and inspiration of too many to name. We would especially like to acknowledge the following:

Wim Van Wassenhove for his contribution of Chapter 2.

Katie Hammon, Kattie Washington, Kiruthika Govindaraju, and other staff at Elsevier for enabling the publication of this book.

The engineers and scientists who have preceded us in developing the technologies that we present.

Our families, friends, and colleagues for their encouragement.

You the readers, who are the ultimate motivation for producing this work.

CHAPTER 1

Introduction to Natural Gas Processing Plants

1.1 INTRODUCTION

For natural gas to be available to the market, it must be gathered, processed, and transported. Natural gas produced from the well contains hydrocarbons, carbon dioxide, hydrogen sulfide, and water together with many other impurities. Raw natural gas after transmission through a network of gathering pipelines therefore must be processed in a safe manner and with minimal environmental effect before it can be moved into long-distance pipeline systems for use by consumers. Although some of the required processing can be accomplished at or near the wellhead (field processing), the complete processing of natural gas takes place at a processing plant, usually located in a natural gas producing region. The objective of a gas processing plant is to separate natural gas, associated hydrocarbon liquids,[1] acid gases, and water from a gas producing well and condition these fluids for sale or disposal. The processing philosophy depends on the type of project being considered and the level of treating required, i.e., the difference between the feed gas and product specifications. This determines what components will need to be removed or recovered from the gas stream.

This chapter describes the scope of natural gas processing and briefly reviews the function and purpose of each of the existing process units within the gas processing plants.

1.2 NATURAL GAS PROCESSING OBJECTIVES

Raw natural gas stream must be treated to comply with emissions regulations and pipeline gas specifications. Typical pipeline gas specifications are shown in Table 1.1. The specifications are to ensure gas qualities and provide a

---

[1] Associated hydrocarbon liquids, known as natural gas liquids (NGLs), include ethane, propane, butane, isobutane, and natural gasoline (pentanes plus).

Modeling, Control, and Optimization of Natural Gas Processing Plants
ISBN 978-0-12-802961-9
http://dx.doi.org/10.1016/B978-0-12-802961-9.00001-2

Table 1.1 Typical pipeline gas specifications

Characteristic	Specification
Water content	4−7 lbm H_2O/MMscf of gas
Hydrogen sulfide content	0.25−1.0 grain/100 scf
Gross heating value	950−1200 Btu/scf
Hydrocarbon dew point	14−40°F at specified pressure
Mercaptan content	0.25−1.0 grain/100 scf
Total sulfur content	0.5−20 grain/100 scf
Carbon dioxide content	2−4 mol%
Oxygen content	0.01 mol% (max)
Nitrogen content	4−5 mol%
Total inerts content ($N_2 + CO_2$)	4−5 mol%
Sand, dust, gums, and free liquid	None
Typical delivery temperature	Ambient
Typical delivery pressure	400−1200 psig

clean and safe fuel gas to the consumers. The product gas must meet the heating values or Wobbe Indexes[2] specifications, which are required to ensure optimum operation of gas turbines and combustion equipment to minimize emissions. Pipeline operators also require the product gas to be interchangeable and similar in properties with existing pipeline gas.

When the gas is high in heavy hydrocarbon contents, they must be removed to meet the heating value specification. The removed natural gas liquids (NGLs) can typically command a higher value than natural gas for the same heating value. Ethane can be used as feedstock to petrochemical plants. Propane and butane can be sold as liquefied petroleum gas (LPG). The C_{5+} components can be exported to refineries as a blending stock for gasoline. The characteristics of various types of NGL products can be found in GPSA Engineering Data Book (2004). The C_{2+} NGL, which is termed "Y-Grade" NGL, shall meet the specifications given in Table 1.2. The Y-Grade liquids must be free from sand, dust, gums, gum-producing substances, oil, glycol, inhibitor, amine, caustics, chlorides, oxygenates, heavy metals, and any other contaminants or additive to the product used to enhance the ability to meet specifications.

Note should be made that sometimes a slight change on the product specifications may have significant impacts on the processing options, which will affect the cost and complexity of the gas processing plant.

---

[2] The Wobbe Index (WI) is calculated by dividing the higher heating value of the gas by the square root of the gas density or molecular weight relative to air. WI is frequently used as a parameter to determine the upper and lower limits of gas composition specified in gas sales or import contracts.

Table 1.2 Y-grade NGL specifications (Mokhatab et al., 2015)

Characteristics	Product specifications
Composition	
Methane, maximum	Not to exceed either 0.5 vol% of total stream or 1.5 vol% of ethane content
Aromatics, maximum	1 wt% in total stream or 10 vol% in contained natural gasoline
Olefins, maximum	1 vol%
Carbon dioxide	500 ppmv or 0.35 liquid volume % of ethane
Corrosiveness	Copper strip at 100°F—1A/1B pass
Total sulfur	150 ppm wt
Distillation: End point at 14.7 psia	375°F maximum
Free water	None at 35°F
Product temperature	60–100°F

1.3 GAS PROCESSING PLANT CONFIGURATIONS

The gas processing plant configuration and complexity depend upon the feed gas compositions and the levels of treating and processing required in meeting product specifications and emission limits. Liquid values can also be the drivers for process complexity, which determines the levels of NGL components to be recovered. Fig. 1.1 shows two simplified gas processing plant schematics. The first scheme is to remove condensate, sulfur, and the heavier components to meet sales gas specifications. The second scheme is to process the feed gas for recovery of the NGL components to increase plant revenues. The residue gas is typically recompressed to a sales gas pipeline. It can also be sent to a natural gas liquefaction plant for liquefied

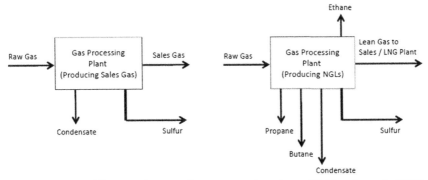

Figure 1.1 Two different schemes of gas processing plants (Mokhatab et al., 2015). *LNG*, liquefied natural gas; *NGL*, natural gas liquid.

natural gas (LNG) production for use as fuel gas to power plants or as a feedstock to petrochemical plants.

There are various technologies, conventional or proprietary, for configuring the gas processing plant. The gas processing plant must be a "fit-for-purpose" design, meeting the project economics and environmental requirements (Mokhatab and Meyer, 2009). Although contaminants and sulfur must be removed to meet emissions requirements as shown in the first scheme, the extent of processing in the second scheme is project specific. The extent of processing depends on the commercial agreements between upstream producers and downstream product distributors and buyers.

1.3.1 Gas Plant with Hydrocarbon Dew Pointing

Raw gas to a gas processing plant can be relatively lean, that is, containing a small amount of C_{2+} hydrocarbons. This lean gas can be processed by the process units as shown in Fig. 1.2. The main process units consist of acid gas removal, gas dehydration, and hydrocarbon dew point control. There are other off-site support systems such as the sulfur recovery with tail gas treating and sulfur production, which are necessary to meet environmental

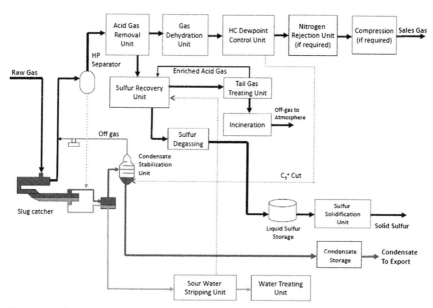

Figure 1.2 Process units in a gas plant with hydrocarbon dew pointing (Mokhatab et al., 2015). HC, hydrocarbon; HP, high pressure.

requirements. If the gas contains liquid condensate, a condensate stabilization unit is required. Nitrogen rejection is required for high levels of nitrogen. Other units, such as gas compression, may also be required.

1.3.1.1 Inlet Separation Facility

Feed gas from production wells arriving at the natural gas processing plant is first separated in the inlet facility, which typically includes slug catchers, separation equipment, and a pressure protection system. Typically a high reliability safety system is installed at the inlet to protect the gas plant from an emergency condition.

The slug catcher captures the largest liquid slugs expected from the upstream operation[3] and then allows them to slowly drain to the downstream processing equipment to prevent overloading the system. The slug catcher design is either a "vessel type" or a "finger type." A vessel-type slug catcher is essentially a knockout vessel (Fig. 1.3). The advantages of the vessel type are that they require significantly less installation area, simple in design and easy to maintain. The traditional finger-type slug catcher consists

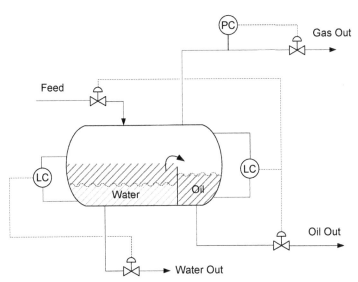

Figure 1.3 Typical vessel-type slug catcher (Mokhatab et al., 2014). *LC*, level controller; *PC*, pressure controller.

---

[3] Some slugs will grow as they travel down the pipeline, whereas others are dampened and disappear before reaching the oil and gas separation facility.

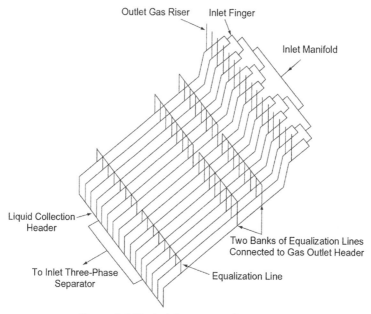

Figure 1.4 Typical finger-type slug catcher.

of multiple long pieces of pipes ("fingers"), which together provide the surge volume (Fig. 1.4). The "finger-type" design is generally less expensive than the vessel type in high-pressure (HP) operation. The disadvantage is the large footprint requirement. It is limited to a land base facility where there is no space constraint.

The vessel-type slug catchers are typically used only if the incoming liquid volumes are fairly stable and relatively small. When the incoming liquid flow rate is uncertain, such as in long gas transmission pipelines, the potentially large liquid volumes would require the use of the finger-type slug catcher (Shell DEP 31.40.10.12−Gen, 1998).

The slug catcher serves as a three-phase separator where the gas, hydrocarbon liquids (condensate), and aqueous phase are separated. The flash gas from the slug catcher is directed to a HP separator to remove any liquid entrainment before entering the gas treatment section. The condensate is processed in the condensate stabilization unit to reduce the vapor pressure to allow storage in atmospheric storage tanks. If the condensate contains mercaptans and other sulfur components, it must be hydrotreated to meet the total sulfur specification, typically 50 ppmw to meet the export requirement.

Sour water separated from the slug catcher is sent to a sour water stripping unit for removal of the acid gas and ammonia contents. The stripped water can be recycled to the process units or further processed in a wastewater treatment system before disposal. The produced water typically contains monoethylene glycol (MEG), which is used for hydrate control in upstream gas transmission pipelines. The MEG is often contaminated with the salts contained in the formation waters. Since salt is nonvolatile, it will remain in the lean glycol during regeneration, which can cause serious corrosion and fouling problems with equipment and pipelines (Son and Wallace, 2000). There are MEG reclamation packages currently available to remove these salts and other contaminants to maintain the required purity. The processed MEG is collected in a lean MEG tank for reinjection to the production field.

1.3.1.2 Condensate Stabilization

The condensate, a hydrocarbon liquid, separated from the slug catchers contains the dissolved light hydrocarbons and H_2S, which must be removed from the liquid to meet the export condensate specifications. A condensate stabilization unit is typically designed to produce a condensate with less than 4 ppm H_2S and Reid vapor pressure specification of 8–12 psi.

A typical condensate stabilization unit is shown in Fig. 1.5. The hydrocarbon liquids are let down to an intermediate pressure in the

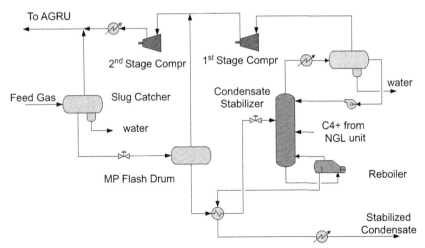

Figure 1.5 Typical condensate stabilization process (Mokhatab et al., 2015). *AGRU*, acid gas removal unit; *MP*, medium pressure; *NGL*, natural gas liquid.

medium–pressure flash drum. The flash gas is compressed back to the feed gas section, and the flashed liquid is further reduced in pressure before entering the condensate stabilizer column. The stabilizer is a fractionator with a reboiler. The overhead vapor from the stabilizer is compressed and recycled back to the feed section. The combined gas stream is sent to the acid gas removal unit (AGRU). The condensate is heat exchanged with the stabilizer feed, cooled, and exported as the stabilized condensate. If significant amounts of mercaptans are present in the feed gas, the condensate requires further treatment, typically by hydrotreating, to meet the product sulfur specification.

The condensate stabilization unit can also be designed to produce an NGL product, as shown in Fig. 1.6. In this configuration, a feed liquid stripper is added before the stabilizer, which removes the C_2 and lighter components. The stripper overhead is recycled back to the feed section, and the bottom is fractionated in the stabilizer into an LPG overhead and a C_{5+} condensate bottom products.

The H_2S content in the condensate from the stabilization unit can typically meet 10 H_2S ppmv specifications. However, it may contain higher levels of organic sulfurs such as carbonyl sulfide (COS) and mercaptans. If condensate is exported as a product, the total sulfur content must be met and a separate hydrotreating unit for removal of the mercaptan content may be necessary. If the condensate is sent to refinery, the condensate can then be blended with the refinery feedstock and treated in the relevant units.

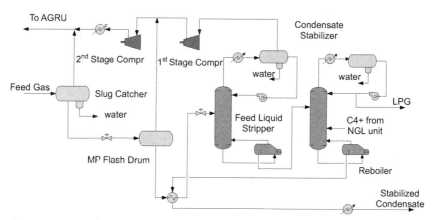

Figure 1.6 Typical condensate stabilization unit with liquefied petroleum gas production (Mokhatab et al., 2015). *AGRU*, acid gas removal unit; *LPG*, liquefied petroleum gas; *MP*, medium pressure; *NGL*, natural gas liquid.

The function of the condensate hydrotreating unit is to remove sulfur compounds from the condensate to meet a desulfurized product specification of 50 ppmw total sulfur. There are several challenges on the design of a hydrotreating unit for the condensate produced from sour gas (Schulte et al., 2009). These include the presence of elemental sulfur, the wide condensate boiling ranges, uncertainties on compositions, and limited data on distribution of sulfur species, aromatic compounds, and metal contaminants.

In the hydrotreater reactor design, various sulfur compounds, such as mercaptans, sulfides, and elemental sulfur are reacted with hydrogen at high temperature and pressure in the presence of a hydrotreater catalyst. The sulfur contents are converted to H_2S, producing a sulfur-free hydrocarbon ($R-S-H + H_2 \rightarrow R-H + H_2S$, where R = Hydrocarbon chain).

Conditions in the reactor are designed to keep the elemental sulfur in solution and avoid deposition of solids on the catalysts. This typically requires a lower operating temperature and higher operating pressure than typically used in naphtha service. Although the purpose of the hydrotreater is desulfurization, some saturation of aromatic compounds also occurs. The configuration of the hydrotreater unit is conventional, similar to a refinery application, and primarily consists of the following sections:

- HP reactor loop—sulfur compounds and elemental sulfur in the condensate are converted to H_2S. The reactor effluent is cooled and product liquids are separated from the recycle gas in a product separator.
- Low-pressure stripping section—the condensate product from the separator is stripped to remove H_2S and H_2.
- Makeup hydrogen compression—hydrogen from the hydrogen plant is compressed to the reactor loop pressure.
- Hydrogen plant—supplies high-purity makeup hydrogen to the HP reactor loop.

1.3.1.3 Acid Gas Removal Unit

The AGRU mainly removes the acidic components such as hydrogen sulfide (H_2S) and carbon dioxide (CO_2) from the feed gas stream to meet sales gas specifications given in Table 1.1. For LNG production, CO_2 must be removed to a level between 50 and 100 ppmv to avoid freezing in the cryogenic exchanger and H_2S must be removed to meet the sales gas specification of 4 ppmv, or $^1/_4$ grains per 100 scf. In addition COS, mercaptans, and other organic sulfur species, which contribute to sulfur emissions must be removed. However, to meet today's stringent sulfur emission requirements, the AGRU alone cannot meet the required specifications and treated gas from these units must be polished with other processes such as molecular sieves, which are specifically designed for removal of the other sulfur components.

A number of processes are available to remove H_2S and CO_2 from natural gas (Stone et al., 1996; Clarke and Sibal, 1998). Some have found wide acceptance in the gas processing industry, whereas others are currently being considered. The selection of an acid gas removal process can have a significant impact on project economics, especially when the feed gas contains a high percentage of acid gas. Carbon dioxide and hydrogen sulfide in the feed gas will also significantly impact the thermal efficiency of the natural gas liquefaction process due to the fact that AGRU is an energy-intensive process.

There are three commonly used solvent absorption processes for acid gas removal in natural gas processing plants: chemical absorption, physical absorption, and the mixed solvents processes (Klinkenbijl et al., 1999). The other processes have limited applications. Membrane separation is only suitable for bulk acid gas removal, whereas the other processing methods, such as cryogenic fractionation and fixed beds adsorption, are not cost-competitive.

1.3.1.3.1 Chemical Solvent Processes

Chemical absorption processes, in which the H_2S, CO_2, and to some extent COS are chemically absorbed, will not remove mercaptans down to low levels due to the low solubility of these components. The advantage of a chemical solvent process such as amine is that the solubility of aromatics and heavy hydrocarbons in the aqueous solvent is low and hence lower hydrocarbon losses. The disadvantage is their high energy consumption, in amine regeneration heat duty and cooling duty.

Common examples of amine processes are aqueous solutions of alkanol amines such as monoethanolamine, diglycolamine (DGA), diethanolamine (DEA), diisopropanolamine (DIPA), and methyldiethanolamine (MDEA). With the exception of MDEA, amines are generally not selective and will remove both CO_2 and H_2S from the gas. Amine can also be formulated by solvent suppliers to increase their selectivity and/or absorption capacity (Hubbard, 2009). Typically, the MDEA selectivity toward H_2S is highest at low operating pressures such as in the tail gas unit, but its selectivity is significantly reduced at high pressure.

When used in treating sour gases to meet the tight CO_2 specification for an LNG plant, the activity of CO_2 absorption is too slow with pure

MDEA, which must be enhanced with a promoter. The most widely used promoted MDEA process is the activated methyldiethanolamine (aMDEA) process, which was originally developed by BASF. The aMDEA process uses piperazine as an activator in MDEA for CO_2 absorption. Since the patent on the use of piperazine with MDEA has expired, the solvent can now be obtained from several amine suppliers such as Dow, Huntsman, and INEOS. The process can also be licensed from technology licensors, such as BASF, UOP, Shell, Lurgi, and Prosernat.

A typical amine process is shown in Fig. 1.7. The feed gas is scrubbed in an amine contactor, which consists of an amine absorption section and a water wash section. The amine absorption section removes the acid gases from the sour feed gas by contacting with a lean amine. The treated gas is washed in the water wash section to recover amine from the treated gas. The water wash section reduces amine makeup requirement and minimizes fouling in the molecular sieve unit.

The rich amine from the amine absorber is flashed in a rich amine flash drum, producing a flash gas that can be used for fuel after being treated with amine. Typically, a lean rich exchanger is used to reduce the regeneration reboiler duty. The amine is regenerated using steam or other heating medium. The lean amine is cooled, pumped, and recycled back to the amine absorber.

The common problem in operating an amine unit is foaming in the amine contactor. Sour gas from the HP separator is at its hydrocarbon dew point, such that the lean amine temperature must be controlled at some

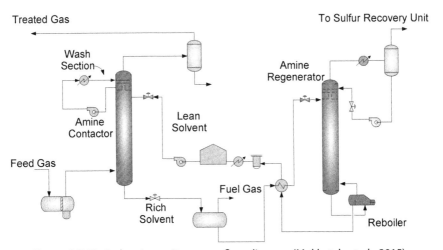

Figure 1.7 Typical amine unit process flow diagram (Mokhatab et al., 2015).

margin above the sour gas temperature, typically at $10°F$, to prevent condensation and subsequent foaming in the absorber. In hot desert areas, where cooling water is not available, process cooling must be by air coolers. In most areas, it is difficult to cool the process gas to below $150°F$. In the amine absorber, removal of the acid gases would increase the gas dew point temperature, which means that the lean amine temperature would need to be further increased. A high lean amine temperature would lower the equilibrium loading of the rich amine, increasing the amine flow rate, making treating difficult. To allow the absorber to operate in hot climate areas, feed gas chilling is required, which can be done with the use of propane refrigeration. The feed gas can be chilled to remove the hydrocarbons, reducing the hydrocarbon dew point, which would allow the amine contactor to operate at a lower temperature, hence reducing the solvent circulation.

1.3.1.3.2 Physical Solvent Processes

Physical absorption processes use a solvent that physically absorbs CO_2, H_2S, and organic sulfur components (COS, CS_2, and mercaptans). Physical solvents can be applied advantageously when the partial pressure of the acid gas components in the feed gas is high, typically greater than 50 psi. Note that the physical solvent acid gas holding capacity increases proportionally with the acid gas partial pressure according to Henry's law and can be competitive to chemical solvent processes because of the higher loading and less heating duty. However, physical solvents are not as aggressive as chemical solvents in deep acid gas removal and may require additional processing steps. Depending on the acid gas contents, a hybrid treating system, such as physical solvent unit coupled with a sulfur scavenger, may be a better choice than a single amine system (Mak et al., 2012).

There are several proven physical solvent processes such as Selexol (licensed by UOP), Fluor Solvent (licensed by Fluor), or Purisol (by Lurgi). The main advantages of physical solvent processes are that the solvent regeneration can be partially achieved by flashing of the solvent to lower pressures, which significantly reduces the heating requirement for regeneration. In some processes, such as the Fluor physical solvent process (Fig. 1.8), no heating is required as the solvent is regenerated by vacuum flashing or by stripping with treated gas or inert gases. When used in treating high-pressure, high—acid gas content gases, greenhouse gas emissions from physical solvent units are significantly lower than the amine units.

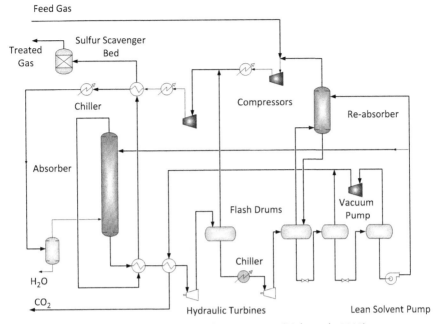

Figure 1.8 Fluor physical solvent process (Mak et al., 2012).

The main disadvantage of the physical solvent unit is the coabsorption of hydrocarbons, which reduces the heating value of the product gas.[4] Unlike an amine unit where the process design is quite straightforward, physical solvent unit designs are more complex. Most of the physical solvent processes are licensed processes and the process configurations and operating conditions vary depending on the licensed solvent used. Typically, physical solvent units require additional equipment, such as recycle compressor, refrigeration, stripper, and flash drums. Gas recycling reduces the hydrocarbon losses, which, however, also increase the solvent circulation. If treated gas is used for stripping, the stripper overhead gas can be recycled back to the absorber, which also increases the solvent circulation.

Therefore the design of a physical solvent unit is more involved and must be carefully optimized to realize the advantages of the physical properties of the solvent.

When treating a HP sour gas, with a high CO_2 to H_2S ratio, coabsorption of CO_2 cannot be avoided. In this situation, a H_2S-selective

---

[4] The high hydrocarbon contents in the acid gas stream entering sulfur recovery unit will cause equipment fouling problems and increase emissions.

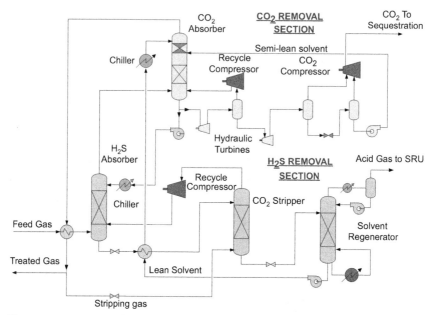

Figure 1.9 Two-stage physical solvent selective process (Mak et al., 2010). *SRU,* Sulfur recovery unit.

solvent such as dimethylpolyethylene glycol (DPEG) can be a viable option. DPEG is marketed by Dow as the Selexol solvent or by Clariant as the Genosorb solvent. The DPEG solvent process can be licensed from UOP as the Selexol process or from Fluor as the EconoSolv process. The process consists of a H_2S removal section and a CO_2 removal section (Fig. 1.9). In the H_2S removal section, the H_2S content is removed from the feed gas using a CO_2-saturated lean solvent from the CO_2 section. To concentrate the H_2S content in the acid gas, the CO_2 content in the solvent is stripped using a slip stream of the treated gas. The stripper overhead gas is recycled back to the H_2S absorber to reduce hydrocarbon losses. This operation effectively reduces the CO_2 content in the rich solvent, hence increasing the H_2S content in the acid gas to the sulfur recovery unit (SRU). The rich solvent is regenerated with steam in the solvent regenerator to produce a lean solvent that is used in the CO_2 section. The CO_2 content in the sulfur-free gas from the H_2S section is removed in the CO_2 absorber. In the CO_2 section, the CO_2-rich solvent is regenerated by flash regeneration by reducing pressure at various levels. The HP CO_2 can be fed directly to the CO_2 compressor, which significantly reduces the compression horsepower for CO_2 sequestration. This configuration can be used to capture over 95% of the carbon content in gasification applications (Mak et al., 2010).

1.3.1.3.3 Mixed Solvent Processes

Mixed solvent processes use a mixture of a chemical and a physical solvent. They are used to treat high—acid gas content gases while meeting the deep removal of the chemical solvents. To some extent, these favorable characteristics make them a good choice for many natural gas treating applications. The Shell Sulfinol process is one of the proven mixed solvent processes.

1.3.1.4 Sulfur Recovery and Handling Unit

Acid gas from the amine regenerator contains concentrated H_2S, which cannot be vented for safety reasons or flared due to acid gas pollution. If reinjection wells are available, acid gas can be reinjected into the reservoirs for sequestration. However, research indicated that the sulfur compounds would have long-term negative impacts on the reservoirs and formation. At present, acid gas is generally processed in an SRU that is closely coupled with a tail gas treating unit (TGTU). With this combination, the sulfur recovery system can meet 99.9% sulfur removal target, which is needed to meet today's emission requirements.

There are many sulfur recovery technologies that are available with different levels of performance in terms of operation and results. The selection of the sulfur technology mainly depends on the amount of H_2S, CO_2, and other contaminants in the feed gas. As a rule of thumb, liquid redox technology is suitable for small SRUs (below 20 tons per day) and for the larger units, the Claus sulfur technology is the most common (Hubbard, 2009).

1.3.1.4.1 Claus Sulfur Recovery Technology

The common method for converting H_2S into elemental sulfur in a gas processing plant is the Claus technology-based process, which is available from several sulfur plant licensors. The Claus process is basically a combustion unit and, to support the sulfur conversion reaction, the acid gas must contain sufficient H_2S to support the heat of combustion. Typically, the H_2S content in the acid gas must be greater than 40 mol%. If the feed gas contains insufficient H_2S, additional processing steps are required, which may require supplemental preheating, oxygen enrichment, or acid gas enrichment (Mokhatab et al., 2015). In addition, if the acid gas contains other contaminants such as ammonia, BTEX (benzene, toluene, ethylbenzene, and xylenes), and mercaptans, a higher combustion temperature is necessary, which may require even higher H_2S content gases.

In a conventional Claus process (Fig. 1.10), the reaction is carried out in two stages. The first stage is the thermal section where air is used to oxidize about one-third of the H_2S content in the sour gas to SO_2. This reaction is highly exothermic and typically about 60—70% of the H_2S in the sour gas is converted to sulfur. In the thermal stage, the hot gases are cooled to 600—800°F and the waste heat is used to generate HP steam. During this process, the S_2 sulfur species are converted to other sulfur species, primarily S_6 and S_8. The gases are finally cooled, to 340—375°F, in a sulfur condenser by generating low-pressure steam.

The second stage is a catalytic stage. The residual H_2S is converted in usually three stages of reactors where sulfur is converted by reaction of the residual H_2S with SO_2 at lower temperatures, typically 400—650°F. The gas must be preheated first to avoid sulfur deposition on the catalyst. Sulfur liquid is condensed and is routed to the sulfur pit for degassing.

Sulfur recovery efficiencies for a two-stage catalytic process are about 90—96%, and for a three-stage process, the efficiencies can be increased to about 95—98%. With an additional selective oxidation stage, such as the SUPERCLAUS process licensed from Jacob Comprimo, sulfur conversion increases to about 99.0%.

1.3.1.4.2 Sulfur Degassing

Liquid sulfur produced from the condensers typically contains 200—350 ppmw H_2S, partially dissolved and partly present in the form of polysulfides. If liquid sulfur is not degassed, H_2S will be released in the storage tanks, which would create toxicity hazard and noxious odor problems. The desirable liquid sulfur product should contain a H_2S concentration of 10 ppmw or less, suitable for transport. To meet this specification, sulfur degasification processes employ a combination of residence time, agitation, and sometimes catalysts. All degassing processes involve agitation of the liquid sulfur and removal of the evolved H_2S with a sweeping gas. Generally, air is used as the sweep gas since oxygen helps to release the H_2S from the polysulfide molecule. The contaminated degassing off-gas is typically vented to the thermal oxidizer for oxidation to SO_2, or directed to the front end of the SRU. A properly sized wire mesh mist eliminator is utilized to minimize potential liquid sulfur entrainment in the overhead vapor stream.

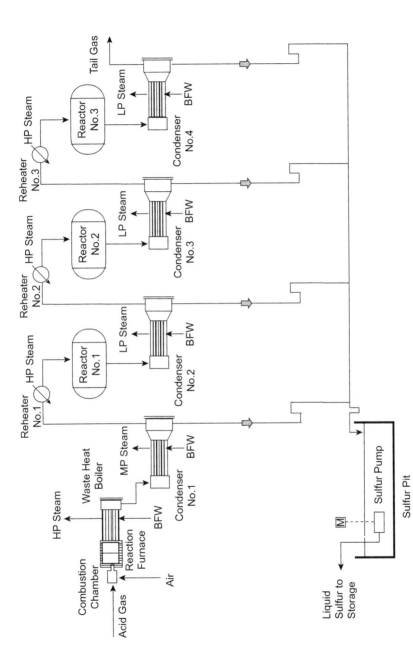

Figure 1.10 Typical three-stage Claus sulfur recovery unit (Mokhatab et al., 2015). *BFW*, boiler feed water; *HP*, high pressure; *LP*, low pressure; *MP*, medium pressure.

The most common degassing processes capable of meeting a 10 ppmw total H_2S specification are the Aquisulf process (licensed from Lurgi) and the D'GAASS process (licensed from Fluor).

1.3.1.4.2.1 Aquisulf Process Among these degassing processes, the earliest process is the Aquisulf process (Nougayrede and Voirin, 1989), with more than 80 references from 15 to 1200 LT/D sulfur (single train). This process is based on two principles: mechanical degassing by agitation and pulverization to promote gas—liquid contact and chemical degassing by catalyst injection to speed up the decomposition of H_2S_X. The Aquisulf process is available in batch and continuous versions. In the continuous version (Fig. 1.11), the liquid sulfur pit consists of two or more compartments. The sulfur in the first compartment is pumped and mixed with the catalyst and sprayed back into the compartment. The liquid sulfur overflows from a weir to the second compartment. The liquid sulfur is again pumped and sprayed in this tank to provide more agitation. The original catalyst was ammonia. However, problems associated with ammonium salts have resulted in the development of an improved version of the proprietary Aquisulf liquid catalyst.

1.3.1.4.2.2 D'GAASS Process The D'GAASS process (developed and commercialized by Fluor in 1996) can achieve 10 ppmw residual H_2S/H_2S_X in liquid sulfur without catalyst addition. There are over 80 D'GAASS units worldwide with a total capacity of over 40,000 LT/D sulfur.

In this process, degasification is carried out in a pressurized vertical vessel using instrument air or clean plant air. In fact, the key to the D'GAASS

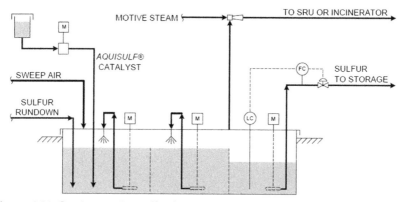

Figure 1.11 Continuous Aquisulf scheme (Nougayrede and Voirin, 1989). *LC,* level controller; *M,* mixer; *SRU,* sulfur recovery unit.

process is the elevated operating pressure typically over 60 psig using the proprietary vessel internals for degassing. Operation at elevated pressure facilitates degassing operation and results in low capital and operating costs. Due to the elevated operating pressure, the D'GAASS contactor is small. The contactor vessel is typically located outside the pit, which makes it easy to install in a grass root unit or for retrofitting an existing facility.

For SRUs with TGTUs required to meet 99.9+% sulfur recovery, conventional degassing systems that vent the vapor stream to the incinerator can be a significant emission source. In these cases, the degassing system is the only emission source from the SRU/TGTU. For the D'GAASS process, 90+% of the H_2S/H_2S_X feed and the overhead stream can be recycled back to the SRU, resulting in a zero emission design.

1.3.1.4.3 Sulfur Storage and Handling

Liquid sulfur, degassed to below 10 ppmw H_2S, has a strong equilibrium driving force to react with air (oxygen) to form SO_2 in air-purged storage tanks (Johnson, 2005). The degassed liquid sulfur, containing traces of SO_2, is typically directed to an insulated and heated carbon steel storage tank to provide a buffer between the SRU and the downstream system. Some H_2S will evolve inside the tank due to the air/sulfur/H_2S equilibrium.

Despite the flammability of H_2S and sulfur in air, the current industry practice is to use an air sweep of the vapor space in the sulfur tank to maintain the H_2S concentration to below the lower explosive limit, which is 3.4% volume at the storage temperature of 330°F (Iyengar et al., 1995).

The stored liquid sulfur can be shipped to the market by tank trucks, railcars, or pipeline. In most cases, liquid sulfur is converted into a solid form for ease of handling and transportation. The liquid sulfur can also be poured into a block for short-term storage during emergency. The sulfur fines reclaimed from the sulfur block can be melted to provide feed for shipments or sulfur forming facilities.

1.3.1.4.4 Tail Gas Treating Unit

To meet today's stringent sulfur emission requirements, use of the Claus or the SUPERCLAUS process alone is not enough as they are limited by chemical equilibrium. To comply with today's emissions requirement, the residual sulfur in the effluent from the Claus unit must be further removed by amine treatment in a TGTU.

Before the tail gas can be processed in the gas treating unit, the SO_2 content must be converted to H_2S, which can then be absorbed by a

H_2S-selective amine. The amine absorbs the H_2S content, which is recycled back to the front section of the Claus unit. This recycle process was first developed by Shell as the SCOT (Shell Claus Off-gas Treating) unit, which has become a standard unit to meet today's emissions requirements. With the use of a selective amine, sulfur recovery of over 99.9% can be met (Harvey and Verloop, 1976). This recycle design can also be integrated with an acid gas enrichment unit as discussed in the following section.

The tail gas unit designs, which are offered by several licensors, typically consist of two sections, the hydrogenation section and the tail gas treating section (Fig. 1.12). In the hydrogenation section, SO_2 is catalytically converted to H_2S in the presence of hydrogen. The conversion of SO_2 must be complete as it will react with amine in the treating unit resulting in amine degradation. The hydrogenated tail gas is then cooled by water in a quench tower. Excess water condensate is removed from the tower and the overhead tail gas is sent to an amine unit.

The tail gas to the amine unit consists of mainly CO_2 and a very low level of H_2S. The TGTU is designed to selectively absorb H_2S while rejecting the CO_2 content before proceeding to a thermal oxidizer or incineration. The H_2S–selective amine can be a formulated MDEA for

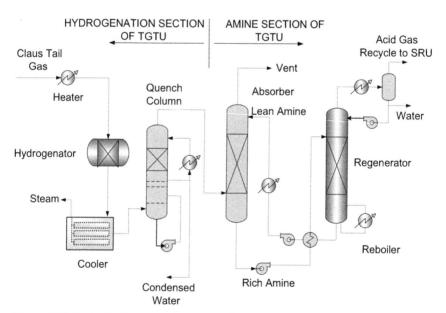

Figure 1.12 Typical tail gas treating and hydrogenation unit (Mokhatab et al., 2015). *SRU*, sulfur recovery unit; *TGTU*, tail gas treating unit.

sulfur removal or sterically hindered amines. The formulated MDEA is available from several amine suppliers, and the sterically hindered amine, such as FLEXSORB, can be licensed from ExxonMobil.

It should be noted that H_2S absorption equilibrium is favored by low amine temperatures, which may not be achievable with air cooling alone, especially in hot climate areas. Typically, a chilled water system or refrigeration is used in the TGTU.

1.3.1.4.5 Acid Gas Enrichment Unit

In a conventional Claus SRU, the acid gas typically contains over 40 mol% H_2S, which would provide sufficient heating value to support the heat of reaction. When a sour gas with high CO_2 content is treated, the acid gas produced from the treating unit will contain a lower H_2S content. To operate the Claus unit, the H_2S content in the acid gas must be concentrated by rejecting its CO_2 content using an acid gas enrichment unit (AGEU).

Similar to the TGTU, the AGEU uses an H_2S-selective solvent such as the formulated MDEA or the sterically hindered amines. There are several patented configurations that can be used to treat lean acid gases. These processes can be used to enrich an acid gas with 10 mol% H_2S to as high as 75% (Wong et al., 2007). In addition to rejection of CO_2, the enrichment unit can also be used to reject hydrocarbons, mercaptans, and other contaminants that are known to be problems with Claus unit operation. These processes are based on recycling a portion of the acid gas from the regenerator back to the absorber, as shown in Fig. 1.13.

1.3.1.4.6 Sulfur Scavenger Unit

With the concerns for global warming, the application of SRU and TGTU may not be sufficient. If less than 1 ppm sulfur specification is required, a sulfur scavenger fixed bed can be used. One of the common sulfur scavenger processes is the PURASPEC process, which can be licensed from Johnson Matthey. PURASPEC has been used as a polishing unit downstream of a gas treating unit. When used in conjunction with a physical solvent unit such as the FLUOR Solvent process, it can be used to treat a wide range of gases with high CO_2 content without the use of heat, hence minimizing greenhouse gas emissions (Mak et al., 2012).

Compared with other sulfur options, sulfur scavenger is expensive. However, it can produce a treated gas with a H_2S content that cannot be economically achieved by other methods. When used by itself alone, the scavenger bed is typically economical for low-sulfur feed gases, typically

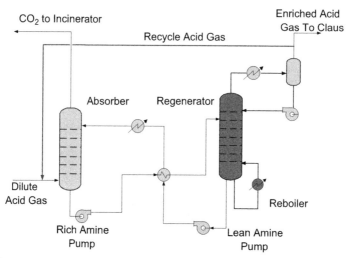

Figure 1.13 Typical acid gas enrichment process (Mokhatab et al., 2015).

below 500 kg/day. For higher sulfur throughput, the sulfur scavenger system can best be used as a polishing unit downstream of a gas treating unit in combination with a Claus unit or Redox unit.

1.3.1.5 Gas Dehydration Unit

Treated gas from the AGRU is fed to the gas dehydration unit to avoid any potential problems resulting from water vapor condensation and accumulation in the pipelines (i.e., plugging, erosion, and corrosion).

There are several methods of dehydrating natural gas, including absorption, adsorption, and direct cooling of the wet gas. However, the absorption processes using liquids (i.e., glycol) and the adsorption processes using solid desiccants (i.e., molecular sieves, silica gels) are the most common. The direct cooling method by expansion or refrigeration, with injection of hydrate inhibitors, is common for less dew point depression in the production of pipeline gas in mild weather regions. Several other advanced dehydration technologies (i.e., membranes and supersonic processes) offer some potential advantages, particularly for offshore applications due to their compact design. However, they have limited commercial experience. There are other solvents that can remove both heavy hydrocarbons and water, including DPEG and methanol, but water removal by these process solvents is considered incidental and typically cannot be customized.

If dehydration is required only to meet the pipeline specification of 4—7 lb/MMscf, any of the above-mentioned processes can be applicable.

However, the typical glycol dehydration process is suitable for meeting pipeline gas specification as low as −40°F, and is more economical than molecular sieve technology. Molecular sieve dehydration processes, which are usually chosen for deep dehydration to meet a low water dew point for NGL recovery or LNG production, will be discussed in detail in Section 1.3.2.2.

Although many liquids possess the ability to absorb water from gas, the liquid that is most desirable to use for commercial dehydration purposes should possess the following properties:
1. High absorption efficiency
2. Easy and economic regeneration
3. Noncorrosive and nontoxic
4. No operational problems, such as high viscosity when used in high concentrations
5. Minimum absorption of hydrocarbons absorption in the gas and no potential contamination by acid gases.

Glycols are the most widely used absorption liquids as they provide the properties that meet the commercial application criteria. Several glycols have been found suitable for commercial application. The commonly available glycol properties can be found in manufacturers' websites. Their pros and cons can be summarized as follows (Katz et al., 1959):
1. MEG: High vapor pressure and seldom used in contactor at ambient temperature due to high losses in the treated gas. Normally, it is used as hydrate inhibitor whereby it can be recovered from gas by separation at below ambient temperatures. It is used in glycol injection exchanger operating at −20°F to minimize losses.
2. Diethylene glycol: High vapor pressure leads to high losses in contactor. Low decomposition temperature requires low reconcentrator temperature (315−340°F) and thus glycol purity is not high enough for most applications.
3. Triethylene glycol (TEG): Relatively low vapor pressure when operating at below 120°F. The glycol can be reconcentrated at 400°F for high purity. Dew point depressions up to 150°F can be achieved with enhanced glycol process like DRIZO.
4. Tetraethylene glycol: More expensive than TEG but less glycol loss at high gas contact temperatures. Reconcentrate at 400−430°F.

TEG is the most common liquid desiccant used in natural gas dehydration. In process design of the TEG dehydration unit, the upstream unit operation must be considered, as the TEG inlet temperature and water saturation will significantly impact the unit performance. For hot climate regions, the feed gas should be cooled to the lowest possible temperature with cooling water (or chilled water). This is necessary to ensure that the feed gas temperature meets the TEG unit's inlet maximum temperature.

1.3.1.5.1 Conventional TEG Dehydration Process

Fig. 1.14 shows the scheme of a typical TEG dehydration unit. As can be seen, wet natural gas is processed in an inlet filter separator to remove liquid hydrocarbons and free water. The separator gas is then fed to the bottom chamber of an absorber where residual liquid is further removed. It should be cautioned that hydrocarbon liquids must be removed as any entrainments will result in fouling of the processing equipment and produce carbon emissions. The separator gas is then contacted counter-currently with TEG, typically in a packed column.

Typically, the liquid loading on the tray is very low due to the low liquid to gas ratio. To avoid liquid maldistribution, structure packing or bubble cap trays should be used.

TEG will absorb the water content and the extent depends on the lean glycol concentration and flow rate. TEG will not absorb heavy

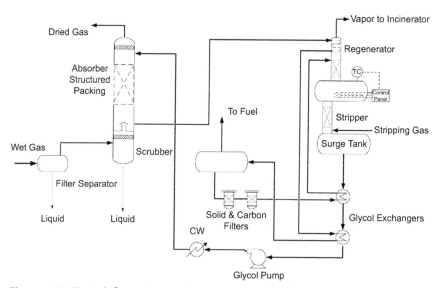

Figure 1.14 Typical flow diagram for conventional triethylene glycol dehydration system (Mokhatab et al., 2015). *CW*, cooling water; *TC*, temperature controller.

hydrocarbons to any degree; however, it will remove a significant portion (up to 20%) of the BTEX components. The BTEX is considered as volatile organic compounds, which must be incinerated to comply with emissions requirements.

Dry natural gas exiting the absorber passes through a demister and sometimes through a filter coalescer to minimize TEG losses. Because of the relatively low TEG flow rate, there is not much sensible heat exchange; hence the dried gas temperature is almost the same as the feed gas.

The rich glycol is used to cool the TEG regenerator overhead, minimizing glycol entrainment and losses from the overhead gas. Rich glycol is further heated by the glycol heat exchanger and then flashed to a tank. The flash gas can be recovered as fuel gas to the facility.

The rich TEG is filtered with solid and carbon filters, heated and fed to the regenerator. The filtration system would prevent pipe scales from plugging the column and hydrocarbons from coking and fouling the reboiler. The water content in the glycol is removed with a reboiler. Heat to the reboiler can be by a fired heater or electrical heater. An electric heater is preferred as it will avoid emission problems, particularly in smaller units. The water vapor and desorbed natural gas are vented from the top of the regenerator.

The dried glycol is then cooled via cross-exchange with rich glycol; it is pumped and cooled in the gas/glycol heat exchanger and returned to the top of the absorber.

1.3.1.5.2 Enhanced TEG Dehydration Process

There are improved regeneration techniques that can produce higher glycol concentration to be used to further lower the water dew point of the treated gas beyond that of the conventional TEG dehydration process. By injecting dry (stripping) gas into the base of the glycol reboiler to further reduce the partial pressure of water, and provide agitation of the glycol in the reboiler, TEG concentration can be increased from 99.1% to 99.6% by weight. Typically, a packed column located below the reboiler section is used for TEG stripping.

DRIZO process (under PROSERNAT license) can regenerate TEG to a higher purity than the conventional gas stripping process. Solvent stripping can produce much higher glycol purities than gas stripping and consequently allows the process to achieve a much larger water dew point depression: up to $-150°F$ and even higher in some cases. The solvent required by the DRIZO process is usually obtained from the C_{6+} (BTEX) present in the natural gas itself and in most cases the process will produce some liquid hydrocarbons.

The main advantages of the DRIZO process are that all BTEX compounds are recovered from the regenerator before being sent to the atmosphere and no external stripping gas is required. The DRIZO technology may be adapted to existing dehydration units, which need to be upgraded to meet requirements for higher glycol purity, or for better emission control of BTEX.

A typical process flow schematic for the DRIZO system is shown in Fig. 1.15. The main difference from the conventional TEG stripping unit is the proprietary separation process in the regenerator overhead where the oil is separated from the aqueous phase. The aqueous phase containing the entrained glycol is refluxed to the regenerator. The hydrocarbon phase is removed, heated, filtered, and used as the stripping gas for TEG regeneration. Purity of the lean glycol can be controlled by adjusting the amount of stripping gas recycle and the regeneration temperature.

There are other solvent stripping processes that can be used to improve the glycol purity, without the use of stripping gas. There are two basic processes that can be used for glycol regeneration. One is the vacuum-based

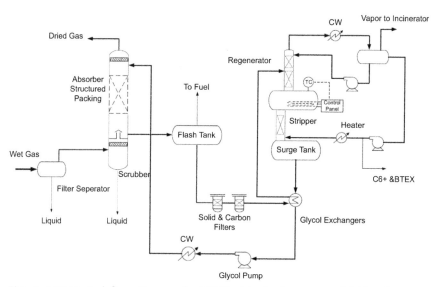

Figure 1.15 Typical flow diagram for DRIZO dehydration system (Mokhatab et al., 2015). *BTEX*, benzene, toluene, ethylbenzene, and xylenes; *CW*, cooling water; *TC*, temperature controller.

Table 1.3 Glycol regeneration process characteristics (GPSA, 2004)

Regeneration process	TEG purity (wt%)	Water dew point depression (°F)
Vacuum	99.2—99.9	100—150
COLDFINGER	99.96	100—150
DRIZO	99.99+	180—220
Stripping gas	99.2—99.98	100—150

process that uses vacuum pressure to reduce the partial pressure of water in the lean glycol. The other approach that can be applied to a glycol regeneration system to accomplish the higher glycol purity is the "Coldfinger" process, which uses a condenser to collect water/hydrocarbons from the reboiler vapor phase and removes them from the reboiler. The Coldfinger process can achieve a TEG concentration of approximately 99.96 wt%. The water depression performance of various TEG dehydration processes are compared in Table 1.3.

1.3.1.6 Hydrocarbon Dew Pointing

The hydrocarbon dew point temperature must be reduced to a temperature that is below the coldest ambient temperature during natural gas transmission. This is to avoid hydrocarbon liquid condensation in the gas transmission pipeline, which is a safety hazard. Depending on the phase envelop of the pipeline gas, the hydrocarbon dew point can actually increase when the pressure is lowered, which must be considered in the design of the unit. Details of the hydrocarbon dew pointing processes are discussed in the following section.

1.3.1.6.1 Hydrocarbon Dew Pointing with Joule—Thomson Cooling

If feed gas is available at high pressure or at supercritical pressure, the gas pressure can be used to generate cooling by isenthalpic expansion. The gas cooling effect will cause heavy hydrocarbon to condense (Fig. 1.16). Glycol injection is typically required to avoid hydrate formation due to the presence of water in the feed gas. Alternatively, the feed gas can be dried using the TEG dehydration system.

As shown in Fig. 1.16, dried gas from the TEG dehydration unit is cooled with the cold separator vapor in a gas/gas shell and tube heat exchanger. The chilled gas is reduced in pressure using a Joule—Thomson (J-T) valve. The J-T letdown operation cools the gas further, producing a liquid condensate in the cold separator.

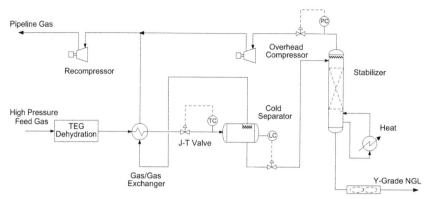

Figure 1.16 Hydrocarbon dew pointing using Joule–Thomson process (Mokhatab et al., 2015). *J-T*, Joule–Thomson; *LC*, level controller; *NGL*, natural gas liquid; *PC*, pressure controller; *TEG*, triethylene glycol.

The pressure of the cold separator is typically maintained between 600 and 700 psig, depending on the feed gas composition. High separator pressure would reduce the recompression horsepower and operating cost. However, high pressure also makes phase separation more difficult, requiring a larger separator. For practical purposes, the separator pressure should stay at 10–20% below the critical pressure of the gas mixture.

The liquid from the cold separator is further processed in a stabilizer, which separates the methane component from the NGL and recompresses it to the sales gas pipeline. The stabilizer is heated with steam or heat medium to produce a Y-grade NGL, which must meet specifications in Table 1.2. The stabilizer typically operates between 100 and 200 psig. The stabilizer pressure is selected based on NGL liquid composition. It is desirable to operate the stabilizer pressure as high as possible to minimize gas compression cost. However, high pressure also increases the bubble point temperature of the stabilizer bottom, which requires a larger reboiler or may also result in thermal degradation. Therefore the stabilizer pressure should be optimized to minimize compression while maintaining a reasonable sized reboiler exchanger.

In many midstream dew pointing units, J-T units are the most common. They can be acquired as low-cost standard modularized units. With adequate pressure drop, relatively low temperatures can be produced for hydrocarbon dew point control. For small gas processing plant operation, J-T is the technology of choice to produce on-spec hydrocarbon dew point gas for sales.

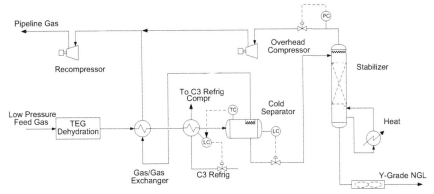

Figure 1.17 Hydrocarbon dew pointing using propane refrigeration (Mokhatab et al., 2015). *NGL*, natural gas liquid; *TEG*, triethylene glycol.

1.3.1.6.2 Hydrocarbon Dew Pointing with Propane Refrigeration

If feed gas is supplied at low pressures, there will not be sufficient pressure to operate a J-T process. In this situation, the feed gas can be chilled at pressure using propane refrigeration. The process configuration, as shown in Fig. 1.17, is similar to the J-T process with the exception that the J-T valve is replaced by a propane chiller. The propane chiller is a kettle-type exchanger with propane evaporating on the shell side. The liquid level on the kettle can be varied to control the cold separator temperature.

The use of propane refrigeration to recover liquids is a more efficient process than the J-T process. The feed gas pressure can be maintained, which would minimize recompression requirement. It is a trade-off between propane compression and feed gas compression, which can be evaluated in an optimization study.

1.3.1.6.3 Deep Hydrocarbon Dew Pointing

J-T units or refrigeration units can be used to meet hydrocarbon dew point specification of the sales gas. Typically, with a low-ethane-content gas, the heating value specification of 1100 Btu/scf can be met. However, with a higher ethane content feed, the heating value of the residue gas may exceed specification. Particularly, in western United States and Europe, a very lean gas with low British thermal unit content is required. California pipelines typically operate with heating values as low as 970 Btu/scf (Mak et al., 2004b), mainly to minimize emissions.

To meet the heating value specifications for these markets, most of the propane and butane components must be removed. Conventional J-T

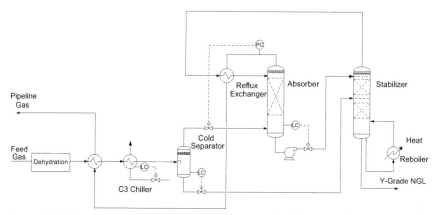

Figure 1.18 Deep hydrocarbon dew pointing process (Mokhatab et al., 2015). *LC*, level controller; *NGL*, natural gas liquid; *PC*, pressure controller.

process or the refrigeration process may not be enough. Consequently, a deeper hydrocarbon dew pointing unit is required, as shown in Fig. 1.18. The process requires both propane refrigeration and J-T expansion. This deep hydrocarbon dew pointing unit requires two columns; an absorber and a demethanizer operating at below −50°F temperature. The feed gas must be dried to below −60°F to avoid hydrate formation in the absorber. Therefore a molecular sieve dehydration or DRIZO dehydration unit must be used instead of the conventional TEG dehydration unit.

The innovation of this process is the reflux exchanger, which uses the absorber overhead vapor to cool and condense the stabilizer overhead vapor, forming a cold reflux stream to the absorber. The reflux system serves two functions. First, it recycles the C_{2+} components to the absorber. Second, the reflux stream condenses the C_{3+} component from the feed gas in the absorber. Consequently, with sufficient J-T cooling, the process can remove most of the propane and heavier components and meet the heating value specification of the sales gas. Propane recovery of over 90% can be achieved with this deep hydrocarbon dew pointing process.

1.3.1.7 Nitrogen Rejection

The nitrogen content in natural gas varies depending on the gas reservoirs. Nitrogen can be naturally occurring in high concentrations in some gas fields, such as in the South China Sea where 30−50% nitrogen content gas can be found. For onshore facilities where nitrogen injection is employed for enhanced oil recovery, nitrogen content can also be very high.

When nitrogen is present in high concentrations, it should be removed to meet the sales gas heating value specification. There are several methods for removing nitrogen from natural gas. Nitrogen removal by cryogenic separation is more efficient than other alternatives. Membrane separators and molecular sieves can be used for nitrogen rejection, but their processing capacity is relatively limited. They are suitable for bulk separation and are not economical to meet stringent specifications. The rejected nitrogen would contain a significant amount of hydrocarbons, which may be an environmental issue.

Several cryogenic schemes are known to reject nitrogen from natural gas with variable nitrogen contents. Each type of these processes can produce a nitrogen waste stream suitable to meet the enhanced oil recovery requirements. The rejected nitrogen stream usually contains a small quantity of hydrocarbon (predominantly methane). If the nitrogen is reinjected, the hydrocarbon contained in the nitrogen is not lost but becomes deferred revenue. If nitrogen is vented, the hydrocarbon content of the nitrogen vent stream must meet the environmental regulation, typically set at 0.5–1 mol% (Millward et al., 2005; Wilkinson and Johnson, 2010). The selection of the nitrogen rejection process must therefore consider the optimum hydrocarbon recoveries, as well as the utility and capital trade-offs. If the unit is installed in an existing facility, the design selection must be suitable to be integrated with the existing equipment.

There are four generic cryogenic processes for nitrogen removal: single-column process, double-column process, preseparation column (or three-column) process, and the two-column process. These processes vary in complexity and efficiency and are discussed individually in the following sections (MacKenzie et al., 2002).

1.3.1.7.1 Single-Column Nitrogen Rejection

The single-column process, shown in Fig. 1.19, utilizes a single distillation column typically operating at 300–400 psig, operated by a closed-loop methane heat-pump system that provides both the reboiler duty and the condensing duty. In this process, feed gas is cooled in heat exchanger HE-1 using the overhead nitrogen and bottom reboiler methane as the coolants. This method, which is applicable for feed gas with nitrogen contents below 30%, can produce HP rejected nitrogen. The drawback is the high power consumption by the heat pump compressor (Kohl and Nielsen, 1997).

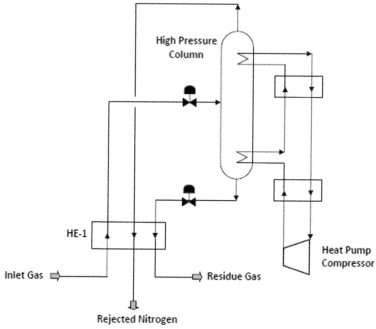

Figure 1.19 Single-column process for nitrogen rejection. *HE*, heat exchanger (Modified after Kohl and Nielsen, 1997).

1.3.1.7.2 Double-Column Nitrogen Rejection

The double-column process is shown in Fig. 1.20. This process uses two distillation columns operating at different pressures that are thermally linked, where the condenser for the HP column is used to reboil the low-pressure column. The process provides all the refrigeration for the separation through the J-T effect by cascaded pressure letdown of the feed.

The nitrogen is produced at low pressure and vented to the atmosphere. The liquid methane product is cryogenically pumped to an intermediate pressure, vaporized in the process, and compressed to pipeline pressure. The process basically fractionates the feed gas stream in the low-pressure column, which operates at the cold portion of the nitrogen rejection unit (NRU), typically at $-250°F$ to $-310°F$, which is prone to CO_2 freezing. The CO_2 content in the feed gas must therefore be removed to a very low level to avoid freezing in the tower.

In this process, which is applicable for the feed gas with nitrogen contents above 30%, the feed gas is cooled and partially condensed in heat exchanger HE-1 using the nitrogen vent stream and methane product and

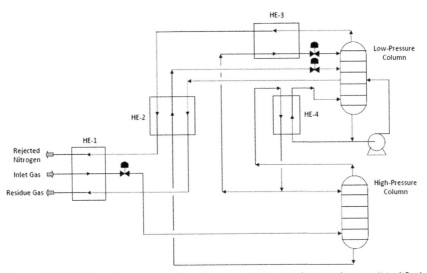

Figure 1.20 Double-column process for nitrogen rejection. *HE*, heat exchanger (Modified after Kohl and Nielsen, 1997).

fed to a HP rectification column that produces a high-purity nitrogen liquid off the top. The bottom product, a mixture of methane and nitrogen, is subcooled in heat exchanger HE-2 and then fed to the low-pressure column, which completes the separation.

The overhead product from the HP column, relatively pure nitrogen, is condensed in HE-4, which also serves as a reboiler for the low-pressure column. A portion of this liquid nitrogen is used as reflux for the HP column, while the remainder is subcooled in heat exchanger HE-3 and used as reflux for the low-pressure column. The methane product from the bottom of the low-pressure column is pumped and then heated to ambient temperature in heat exchangers HE-2 and HE-1 (Kohl and Nielsen, 1997).

1.3.1.7.3 Three-Column Nitrogen Rejection

The three-column (preseparation column) process, as shown in Fig. 1.21, is a variation of the double-column process where the process is made up of the HP column (prefractionator), intermediate-pressure column, and the low-pressure column. The prefractionator can remove the bulk of the methane and CO_2 content as a bottom product, thereby concentrating the nitrogen content in the overhead. With a lower CO_2 content to the cold section, the process is more CO_2 tolerant and can operate with a higher CO_2 content feed gas with CO_2 content up to 1.5 mol%.

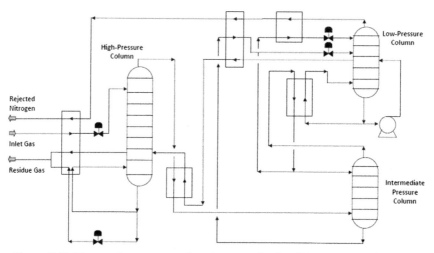

Figure 1.21 Three-column process for nitrogen rejection (MacKenzie et al., 2002).

The prefractionator column operates at a higher pressure with temperature ranging between $-150°F$ and $-180°F$, which would avoid any CO_2 freezing problem. Removing the bulk of the CO_2 from the prefractionator column greatly reduces the CO_2 content in the cold end of the NRU, increasing the process tolerance to CO_2. This was a very important consideration in the design selection since the streams coming from the NGL plants are already dry and treated for CO_2. Additional CO_2 removal would add to the capital and operating cost of the project.

In addition, the heavy hydrocarbons in the feed are recovered in the residue stream from the bottoms of the prefractionator column. This increases the hydrocarbon recoveries and revenues from the NRU.

1.3.1.7.4 Two-Column Nitrogen Rejection

The two-column process, as shown in Fig. 1.22, is similar to the three-column process, without the intermediate column. The process comprises a HP prefractionator and a low-pressure column. Similar to the three column design, the prefractionator reduces the CO_2 content in the feed gas to the low-pressure column and is therefore more CO_2 tolerant.

The design of the two-column process is simpler than the three-column process. If it is used to process a lower-nitrogen-content gas (below 50%), the operating pressure of the low-pressure column can be increased to reduce energy consumption (MacKenzie et al., 2002).

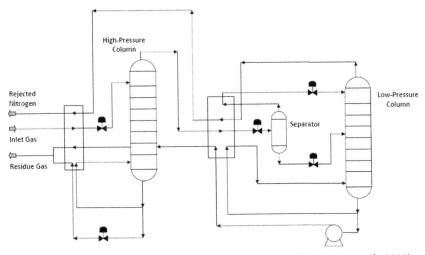

Figure 1.22 Two-column process for nitrogen rejection (MacKenzie et al., 2002).

Process selection for the NRU should be based on operating flexibility, complexity, and sensitivity to feed gas compositions in addition to life cycle costs. The key parameters for process selection are feed gas nitrogen and CO_2 contents, feed pressure, flow rate, methane recovery, and contaminant levels. The more important parameter is the CO_2 tolerance of the selected process (Trautmann et al., 2000). A process that has very little CO_2 tolerance may require a costly deep CO_2 removal system, such as molecular sieve, whereas a more CO_2-tolerant process may only require an amine system. A more CO_2-tolerant process is also more reliable, since it can handle CO_2 removal upsets and avoid shutdown required for derimming due to CO_2 freezing problem.

1.3.1.8 Gas Compression and Transmission

Feed gas to the gas plant is typically reduced in pressure such that phase separation is feasible. Most often, recompression of the residual gas to the pipeline pressure is necessary. Major projects are being planned to move massive amounts of HP sales gas from processing plants to distribution systems and large industrial users through large-diameter buried pipelines. These pipelines utilize a series of compressor stations along the pipeline to move the gas over long distances.

Compressor stations comprise one or more compressor units (designed with enough horsepower and throughput capacity to meet contractual requirements levied on the system), each of which will include a

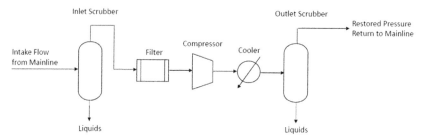

Figure 1.23 Schematic process diagram of single-stage gas compressor station.

compressor and driver together with valves, control systems and exhaust ducts, and noise attenuation systems. Most compressors are powered by natural gas taken directly from the pipeline, but electric-powered compressors are becoming more common.

A typical single-stage compressor station design, as shown in Fig. 1.23, may consist of an inlet scrubber to collect entrained liquids (i.e., water, corrosion inhibitors, and hydrocarbon liquids that may have formed in the gas transmission pipeline) followed by a coalescing filter[5] to remove fine solids (i.e., pipe scale) and hydrocarbon mist from the gas that could otherwise contribute to compressor failure. From the scrubber, the gas is taken to the compressor unit(s) where it is compressed. At the compressor station discharge or between compressor units in case of series arrangement the gas is cooled, typically with an air cooler, and then it passes through a scrubber allowing drainage of any formed liquid. In case of reciprocating compressor usage, a coalescing filter shall be used after the scrubber to remove lube oil mist before introducing the gas into the pipeline.

Each compressor station will be built up from the same functional blocks of equipment shown in Fig. 1.23. Each functional element plays a part in the work of the station and the design and sizing of each is essential to the efficient and safe operation of the plant.

The functional elements include gas scrubbing and liquid removal, compressor and driver units, aftercoolers, pipes, and valves. Controls, including the Supervisory Control and Data Acquisition (SCADA) system, monitoring and data recording, alarms and shut down procedures, and both routine and emergency, are an integral part of the station. Provision also has to be made for venting the compressor and driver housing and buildings, complete with ventilation and fire protection, and safety equipment.

---

[5] A coalescing filter also cleans the gas in each fuel supply to the turbines and gas engines.

1.3.2 Gas Plant for NGL Production

When the feed gas contains a significant amount of hydrocarbon liquids (C_{3+} hydrocarbons), there are economic incentives to produce the LPG and sometimes ethane liquid as by-products. The liquid facility typically includes storage, pipeline, metering, and custody transfer, and must include a safety system to protect against liquid leakage or spillage. This type of plant is complex and costs more than the simple hydrocarbon dew point plant.

Fig. 1.24 illustrates a block flow diagram of a gas processing plant for NGL production. The following sections describe the units that are unique to NGL production. The balance of the plant is similar to the hydrocarbon dew pointing plant.

1.3.2.1 Carbon Dioxide Removal

Carbon dioxide (CO_2) removal is required to meet the sales gas CO_2 specification, typically limited to $2-3$ mol%. CO_2 may need to be removed to even a lower level to avoid CO_2 freezing in the cold section of the NGL recovery unit. Typically, in the propane recovery process, 2 mol% of CO_2 can be tolerated as the NGL column operates at a warmer temperature. Deep CO_2 removal may not be required unless the gas is sent to a

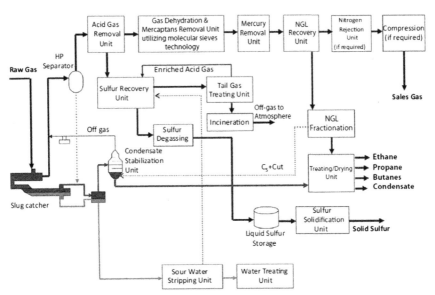

Figure 1.24 Process units in a gas plant for natural gas liquid production (Mokhatab et al., 2015). *HP*, high pressure; *NGL*, natural gas liquid.

liquefaction unit, which, in this case, would require CO_2 to be removed down to 50 ppmv.

If the NGL recovery unit is used for ethane recovery, the demethanizer column would operate at a much lower temperature, which is prone to CO_2 freezing. Even if CO_2 freezing is not a problem, a good portion of the CO_2 will condense with the ethane product, which may not meet the CO_2 specification in the ethane product (typical Y-Grade NGL limits CO_2 content in ethane to 500 ppmv). When ethane recovery is required, design must ensure that CO_2 in the feed gas be removed sufficiently to avoid CO_2 freezing as well as meeting the CO_2 specification of the ethane product.

1.3.2.2 Dehydration and Mercaptan Removal

For NGL recovery, the deethanizer or demethanizer must operate at low temperatures. This requires sufficient water to be removed to avoid hydrate formation in the columns. If only propane recovery is considered, the column operates at a warmer temperature, at about $-60°F$. In this case, the use of DRIZO TEG dehydration may be sufficient. The TEG unit is more compact than the molecular sieve unit and is more suitable for offshore design. If ethane recovery is required, then molecular sieve dehydration is necessary.

Molecular sieves are the only choice for natural gas dehydration to cryogenic processing standards (less than 0.1 ppmv water or $-150°F$ dew point). Molecular sieves can also provide a solution for removal of carbon dioxide and sulfur compounds such as hydrogen sulfide, mercaptans, COS, and other sulfides (with the exception of carbon disulfide[6]) from natural gas and NGLs to very low outlet specifications, either as a stand-alone unit or as a polishing unit within a combination of gas treating processes.

The molecular sieve unit can be designed to remove water to 0.1 ppmv and most mercaptans to 2–3 ppmv. When molecular sieves are used for mercaptan removal, water must be removed first before the mercaptan removal bed. Removal of water and mercaptans on molecular sieves can be installed in a single vessel, where the first layers of molecular sieves remove water, and the subsequent layers of molecular sieves remove mercaptans. Often, 4A, 5A, and 13X molecular sieves are used for the removal of water, light mercaptans, and heavy mercaptans, respectively. Additional layers of specific adsorbents may also be used to remove traces of mercury and to

---

[6] As carbon and sulfur have the same electronegativity, carbon disulfide cannot be polarized and therefore is not adsorbed by molecular sieves.

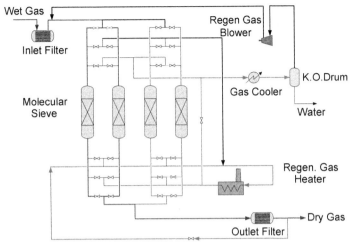

Figure 1.25 Typical molecular sieve dehydration unit (Mokhatab et al., 2015). *K.O*, knock-out.

protect the molecular sieves from plant upsets and unexpected contaminants (Northrop and Sundaram, 2008).

Design of the dehydration and mercaptan adsorption unit is based on the number of fixed beds in a parallel lineup, as shown in Fig. 1.25. In a typical operation, four molecular sieve beds are used, with two of the adsorbers on drying the sweet gas from AGRU, one adsorber thermally regenerated by desorbing the compounds, and the other adsorber cooled before being returned to the next cycle. Each molecular sieve adsorber is typically in adsorption mode for 6 h, followed by 3 h of heating and 3 h of cooling. Every 12 h, the cycle returns to the same point and repeats.

The cyclic operation of the molecular sieve unit imposes stress on the sieve materials requiring bed replacement every 3–5 years. The operation conditions of the cycles, such as flow rates, pressures and temperatures, and regeneration conditions should be evaluated in a dynamic model that can be used to optimize the dehydration operation and extend the bed life (Mock et al., 2008).

If the solvent used in the AGRU is a physical solvent or mixed chemical/physical solvent, a significant portion of the mercaptans may have already been removed in the AGRU, thereby reducing the molecular sieve mercaptan removal requirement. On the other hand, chemical solvents such as DEA, DGA, or MDEA (accelerated) do not absorb any appreciable amount of heavy hydrocarbons and mercaptans, which must be handled by the downstream molecular sieve unit. When the molecular sieves are regenerated, the wet molecular sieve regeneration gas containing the mercaptans is returned to the AGRU inlet. The mercaptan concentration in the gas entering the AGRU will build up until the amount removed in the AGRU equals the incoming mercaptans.

The split of the mercaptan removal between the ARGU and the molecular sieves should consider factors such as operating flexibility on different gas compositions and flow rates, environmental and emissions, and operating and capital costs (Klinkenbijl et al., 1999; Grootjans, 2005). When using the molecular sieves for both dehydration and mercaptan removal, the amount of mercaptans must be purged from the system by one of several means:

• Use the wet regeneration gas as fuel gas. This is unlikely as the regeneration gas typically far exceeds the plant's fuel gas requirement. Also the sulfur oxides from incineration of the mercaptans may exceed the plant's sulfur emissions limit.

• Add a physical solvent AGRU to treat the regeneration gas.

The regeneration gas from the AGRU is processed in the SRU. If the amine unit is used as the AGRU, the majority of the mercaptans will come from the molecular sieve beds during regeneration. The mercaptan release will vary during the regeneration cycle and will peak at some point. This has an impact on the SRU as the SRU catalyst is designed for a certain mercaptan level and is very sensitive to mercaptan load changes (Carlsson et al., 2007; Bradley et al., 2009). To avoid fouling of the SRU catalysts from excessive mercaptan loads, some leveling facilities should be provided to even out the mercaptan flow to the SRU.

The branched-type mercaptans that slip through the AGRU are usually not caught by the molecular sieves and are left in feed gas to the downstream unit. Since these branched-type mercaptans have volatility close to pentane, they are removed together with the NGL stream and will concentrate in the pentane plus condensate (Mokhatab and Meyer, 2009; Mokhatab and Hawes, 2015). To meet the sulfur specification in the condensate product, further processing such as hydrotreating is required (Mokhatab et al., 2015).

1.3.2.3 Mercury Removal

Almost every LNG plant will have a mercury removal unit installed. This is because the consequence for the LNG plant from mercury attack is severe and because it is difficult to predict the mercury contents from production reservoirs. Low levels of mercury can result in severe corrosion of the brazed-aluminum heat exchangers used in cryogenic systems, and can potentially pose environmental and safety hazards. The presence of mercury in the feed stocks to petrochemical plants will also cause poisoning of precious metal catalysts (Carnell and Row, 2007). For this reason, the LNG plant is designed with a conservative design that requires mercury removal to levels below 0.01 $\mu g/Nm^3$.

Most of the current methods for removing mercury from natural gas and hydrocarbon liquids use fixed beds of mercury removal materials. The fluid flows through the fixed bed in which mercury reacts with the reactive reagent in the mercury removal vessel producing a mercury-free product (Kidnay and Parrish, 2006).

There are two types of mercury removal materials: nonregenerative mercury sorbents and regenerative mercury adsorbents.

1.3.2.3.1 Nonregenerative Mercury Sorbents

In nonregenerative mercury removal process, the mercury reacts with the sulfur to form a stable compound on the sorbent surface. A number of different mercury removal sorbents are available with various tolerances to operating temperature, liquid hydrocarbons, and water (Markovs and Clark, 2005). The use of sulfur-impregnated activated carbon is a proven commercial process for removing the mercury.[7] However, there are drawbacks to this method of mercury removal, where sulfur-impregnated carbon can only be used with dry gas since it has a high surface area and small pore size. This also restricts the access of mercury to the sulfur sites and increases the length of the reaction zone. In addition, sulfur can be lost by sublimation and dissolution in hydrocarbon liquids. This again reduces mercury removal capacity. Furthermore, it is often difficult to dispose of the spent mercury-laden carbon material (Abbott and Oppenshaw, 2002).

Recognition of these problems has led to examining technologies other than sulfided carbon, where a range of nonregenerable absorbents utilizing transition metal oxides and sulfides instead of carbon have been developed to improve on existing mercury removal technologies, in which the discharge absorbents can be safely handled. In these systems, the reactive metal is incorporated in an inorganic support and the absorbent is supplied with reactive sulfide component by either ex situ or in situ

---

[7] Sulfur is the active ingredient, securely fixing mercury as sulfide in the microporous structure of carbon.

sulfiding (Carnell and Row, 2007). Johnson Matthey has taken this concept further and supplied an established range of absorbents marketed under the PURASPEC brand. The PURASPEC materials are a mixture of copper sulfide/copper carbonate, zinc sulfide/zinc carbonate and aluminum oxide, which can operate in a wet gas environment (Row and Humphrys, 2011).

Note that the nonregenerative methods appear to be simple since no regeneration equipment and special valving are required. However, disposal of the used sorbent can be a problem since the sorbent not only picks up the mercury, but it will often contain other hazardous material such as benzene and other hydrocarbons (Markovs and Clark, 2005).

1.3.2.3.2 Regenerative Mercury Adsorbents

The regenerative mercury removal process utilizes silver on molecular sieve (such as UOP's HgSIV sieve) to chemisorb elemental mercury. The mercury saturated bed is then regenerated by hot regeneration gas. The regeneration gas is typically heated to 550°F. The mercury can later be recovered in the condensed water. In this method of mercury removal, since the mercury does not accumulate on the adsorbent, it would avoid the disposal problems with spent adsorbent. However, there will be trace of mercury left in the regeneration gas, which can be removed with a non-regenerative mercury bed (Markovs and Clark, 2005).

1.3.2.3.3 Process Considerations

There are four possible options for mercury removal (Mokhatab et al., 2015):

Option 1: Installing nonregenerative mercury removal sorbents at the plant inlet. This option removes all the mercury and ensures no mercury contamination in the rest of the plant (Edmonds et al., 1996). However, the large volume of feed gas and acid gases require a large removal system, which may be challenging in design and operation.

Option 2: Installing a nonregenerative mercury removal sorbent down-stream of the AGRU, just before the molecular sieve unit. This option reduces the size of the beds to some extent, but it poses the risks of mer-cury contamination in the AGRU solvent system.

Option 3: Add a silver-impregnated mercury sieve section to the molec-ular sieve beds. Although this option can remove water, mercaptans, and mercury at the same time, and avoid the need of a separate mercury bed, it presents problem with high mercury content in the regeneration water that would pose operating hazards unless treated by another mercury removal step (Hudson, 2010).

Option 4: Installing a nonregenerative mercury removal bed or a silver-impregnated molecular sieve bed after the molecular sieve unit. Although this option yields good mercury removal performance as the feed gas is dry and clean, it cannot avoid the mercury contamination problem in the AGRU and molecular sieves upstream and may pose operating hazards (Eckersley, 2010).

Life cycle costs, adsorbent disposal methods, mercury levels, environmental limits, operating hazards, and plant operator procedure must be evaluated in the selection of a suitable mercury removal system. The optimum mercury removal method can also be a combination of nonregenerative and regenerative mercury removal system (Markovs and Clark, 2005).

1.3.2.4 NGL Recovery

Recovery of NGL components in gas not only may be required for hydrocarbon dew point control in a natural gas stream (to avoid the unsafe formation of a liquid phase during transport), but also yields a source of revenue. Regardless of the economic incentive, gas usually must be processed to meet the specification for safe delivery and combustion. Hence, NGL recovery profitability is not the only factor in determining the degree of NGL extraction.

The NGL recovery unit can be designed for propane recovery or ethane recovery. For operating flexibility, the NGL recovery process can be designed for either ethane recovery or ethane rejection when ethane margins are low. Another alternative is to design the unit for propane recovery that can be operated on ethane recovery. More details of the various NGL recovery processes are discussed in the following section.

1.3.2.4.1 Lean Oil Absorption

The lean oil absorption process was developed in the early 1910s and was used exclusively until the 1970s. The absorption unit uses a lean oil to absorb the C_{3+} components, followed by a deethanizer, and a rich oil still to regenerate the rich oil. Propane and butane products can be produced. A typical refrigerated lean oil absorption process is shown in Fig. 1.26.

To allow the unit to operate at low temperatures, the feed gas must be injected with ethylene glycol solution to avoid hydrate formation in the heat exchangers. The feed gas is cooled by propane refrigeration and

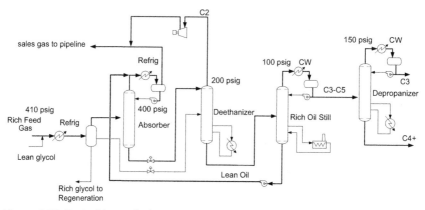

Figure 1.26 Typical lean oil absorption process (Mokhatab et al., 2015). *CW*, cooling water.

separated in a cold separator, typically at about 0°F. The separator liquid is sent to the deethanizer while the separator vapor is routed to the absorber operating at about 400 psig. Refrigerated lean oil is used to absorb the C_{3+} content from the feed gas, producing a lean gas and a propane-rich bottom, which is sent to the deethanizer. The deethanizer operates at a lower pressure, typically at 200 psig, producing an ethane-rich gas and a rich oil bottom containing the C_{3+} components.

The deethanizer overhead is compressed to the sales gas pipeline or used as fuel gas. The bottom is further processed in a rich oil still, which regenerates a lean oil to be recycled back to the absorber, and an overhead distillate containing the C_{3+} components. The C_{3+} stream can be fractionated in a depropanizer, which produces the propane and butane product. Because of the high boiling material of the lean oil, a fired heater is used in the rich oil still. If necessary, the lean oil composition can be controlled using a lean oil still (not shown) to remove the heavy tails of the lean oil from the process.

A typical refrigerated lean oil process can achieve 50–60% propane recovery, depending on the feed gas composition. Because of the high equipment counts and the process complexity, lean oil absorption processes are not cost competitive to expander plants and are seldom used today.

1.3.2.4.2 Turboexpander NGL Recovery Processes

The term "turboexpander" refers to an expander/compressor machine as a single unit. It consists of two primary components, the radial inflow expansion turbine and a centrifugal compressor integrated as a single assembly. The expansion turbine is the power unit and the compressor is the driven unit.

In cryogenic NGL recovery processes, the turboexpander achieves two different but complementary functions. The main function is to generate refrigeration to cool the gas stream. This is done by the expansion turbine end that expands the gas isentropically by extracting the enthalpy from the gas stream, causing it to cool. The other function is the use of the extracted energy to rotate the shaft to drive the compressor end of the turboexpander, which recompresses the residue gas stream.

The first turboexpander unit was built in 1964 for NGL recovery in the city of San Antonio, Texas. The gas is supplied at 700 psig pressure and is let down in pressure to about 300 psig to the demethanizer. Methanol injection was used for hydrate inhibition. Until this time, LPG recovery was mainly achieved with refrigerated lean oil, which is described in a later section.

The first turboexpander process patent was issued to Bucklin (Fluor) in 1966. The flow schematic is shown in Fig. 1.27. The concept was to use a turboexpander to generate cooling instead of the J-T valve, which is a less efficient method of cooling. The feed gas is cooled by the cold demethanizer overhead, and separated in the HP cold separator. The separator vapor is let down in pressure using the turboexpander and fed to the top of the demethanizer as a reflux. The HP cold separator liquid is let down in

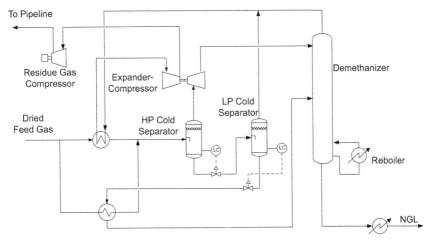

Figure 1.27 Process flow schematic of first turboexpander natural gas liquid recovery process (Bucklin, 1966). *HP*, high pressure; *LC*, level controller; *LP*, low pressure; *NGL*, natural gas liquid.

pressure to the low-pressure cold separator. The separator liquid is further let down and used to cool the feed gas before it is fed to the lower part of the demethanizer. The demethanizer bottom product is heated with steam to control the methane content (see Table 1.2).

Earlier NGL recovery units did not have the benefits of the compact brazed aluminum heat exchanger and required multiple shell and tube heat exchangers to achieve the chilling requirements. Even with an extensive heat integrated scheme, high NGL recovery could not be achieved. The other limitation of earlier expander designs was the low expansion ratio, which required two expanders operating in series to achieve a high expansion ratio. A higher expansion ratio would generate more cooling that was necessary for the high NGL recovery.

The process flow schematic of a typical turboexpander NGL recovery unit built in the 1970s is shown in Fig. 1.28. The process uses two expander—compressor sets operating in series, and several separators to produce various liquid streams that are fed to different locations in the demethanizer. The process also requires propane refrigeration for feed gas cooling. The NGL recovery process could achieve about 70% ethane recovery and was typically based on a relatively rich feed at about 950 psig pressure. Most of these NGL recovery plants are still operating today, but they are processing much leaner gas, as the reservoirs deplete. The leaner gas has an impact on the NGL recovery level. The heat exchanger surface areas are no longer adequate, and the expanders are operating at lower efficiency. Consequently, ethane recovery drops to about 55%.

The main contributors to the success of today's NGL recovery plants are the turboexpanders and the brazed aluminum exchangers. The application

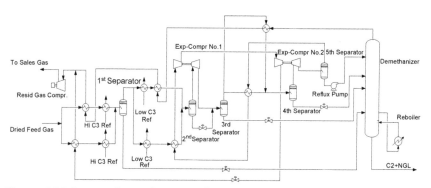

Figure 1.28 Process flow schematic of earlier turboexpander natural gas liquid recovery unit (Mokhatab et al., 2015). *NGL*, natural gas liquid.

of turboexpanders to the natural gas industry began in the early 1960s, which was followed by the development of brazed aluminum heat exchangers.

1.3.2.4.3 Modern NGL Recovery Processes

Modern NGL recovery processes are based on turboexpanders using various reflux configurations. There are many patented processes that can be used to improve NGL recovery, either for propane recovery or ethane recovery. In addition, some of the ethane recovery units can operate in ethane rejection mode with minimum loss in propane. Similarly, some of the propane recovery units also operate on ethane recovery. These processes are discussed in detail in the following sections.

1.3.2.4.3.1 Dual-Column Reflux Process The dual-column process was quite common among NGL recovery units. Typically, the first column acts as an absorber recovering the bulk of the NGL components and the second column serves as deethanizer during propane recovery and demethanizer during ethane recovery. Adding refluxes to the dual-column design process was originally configured for high propane recovery. The process is very efficient and can achieve over 99% propane recovery. The process flow schematic of the dual-column reflux process is shown in Fig. 1.29.

Feed gas, typically supplied at about 1000 psig, is first dried using molecular sieve dryers, and then cooled in a feed exchanger Brazed

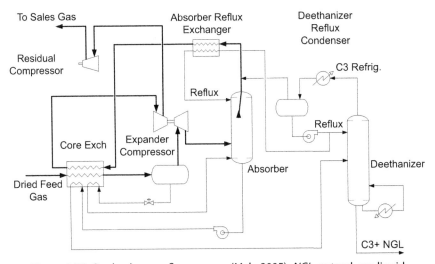

Figure 1.29 Dual-column reflux process (Mak, 2005). *NGL*, natural gas liquid.

Aluminium Heat Exchanger (BAHX). The feed gas is chilled with the absorber overhead vapor, cold separator liquid, and the absorber bottoms. Refrigeration is generated using turboexpander by expanding the separator gas to the absorber typically operating at 450 psig. When processing a rich feed gas, propane refrigeration is often used to supplement the cooling requirement.

The absorber bottoms liquid is pumped by the bottoms pump to the deethanizer, which operates at a slightly higher pressure than the absorber. The deethanizer overhead is refluxed using propane refrigeration, producing a C_2-rich liquid that is used as reflux to the deethanizer and the absorber. The absorber overhead is used to subcool the absorber reflux stream, which further enhances the separation process. With the dual–reflux process, a very high propane recovery can be achieved with very low energy consumption.

1.3.2.4.3.2 Gas Subcooled Process
The gas subcooled process (GSP) is common for ethane recovery in the gas processing industry. The process was invented by Ortloff in the late 1970s. The schematic of a typical GSP process is shown in Fig. 1.30. The unique feature of the GSP process is the split flow arrangement using a portion of cold separator vapor (about 30%) to form a lean reflux stream to the demethanizer. The refrigeration content

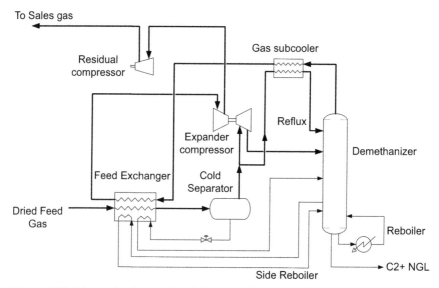

Figure 1.30 Schematic of gas subcooled process (Pitman et al., 1998). *NGL*, natural gas liquid.

of the residue gas from the demethanizer is used in subcooling the reflux stream in the gas subcooler. The remaining portion is expanded in a turboexpander, generating the low-temperature refrigeration for the demethanizer. One or two side reboilers on the demethanizer are typically used to reduce the refrigeration requirement.

The demethanizer reboiler is used to produce an ethane-rich liquid, which must meet methane and CO_2 specifications (see Table 1.2). The column typically operates between 350 and 450 psig, depending on the feed gas inlet pressure, the feed gas composition, and the ethane recovery level.

1.3.2.4.3.3 Ortloff SCORE The SCORE (single-column overhead recycle process) licensed by Ortloff Engineers, Ltd., is based on the same reflux principle as the dual column reflux process. The SCORE process is designed to recover over 99% propane from the feed gas in a single-column configuration.

This process recovers the C_{3+} hydrocarbons from the feed gas and produces a lean residual gas for sales. Alternatively, the residue gas can be sent to the natural gas liquefaction plant. It has been used in natural gas liquefaction units to accomplish the function of the scrub column for the removal of C_{5+} hydrocarbons. The recovered NGL can be exported or used for blending with LNG if a higher heating value LNG is required. Fig. 1.31 shows a simplified process schematic of a typical SCORE process (Thompson et al., 2004).

1.3.2.4.3.4 Residue Gas Recycle When high ethane recovery is required, additional cooling is required by recycling a portion of the residue gas as reflux to the absorber. The process flow schematic of a residual gas recycle process is shown in Fig. 1.32. With this configuration, C_2 recovery, as high as 95%, can be achieved.

Chilling can be produced by lowering the column pressure or increasing the recycle gas rate. However, if the feed gas contains CO_2, a lower demethanizer pressure would lower the column temperature, which is more prone to CO_2 freezing. Lowering demethanizer pressure also requires more residue gas compression. Typically, the optimum pressure of the demethanizer column falls between 350 and 450 psig.

Optimum C_2 recovery levels should be evaluated based on the project economics. Typically, a C_2 recovery level greater than 95% would require a significant increase in gas recycle and equipment cost, which may not be economically justified.

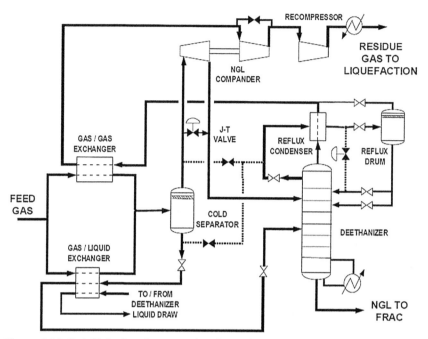

Figure 1.31 Ortloff single-column overhead recycle process (Thompson et al., 2004). *FRAC*, fractionation; *J-T*, Joule–Thomson; *NGL*, natural gas liquid.

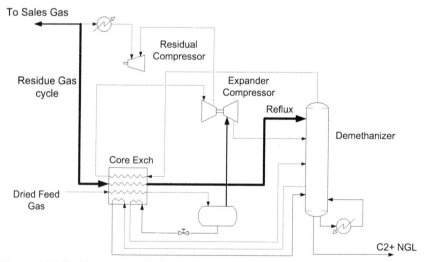

Figure 1.32 Residue gas recycle for high ethane recovery (Mokhatab et al., 2015). *NGL*, natural gas liquid.

1.3.2.4.3.5 Fluor Twin-Column High Absorption Process When the sales gas must be compressed to the pipeline pressure, it is desirable to operate the demethanizer at as high a pressure as possible. However, this may not be feasible in conventional demethanizer designs, such as that shown in Fig. 1.30.

Conventional demethanizer is a single-column design, which operates between 350 and 450 psig. The operating pressure should be well below the critical pressure. More importantly, the critical region of the demethanizer column is the column bottom, which has the highest temperature. Increasing the column pressure will increase the bottom temperature, further moving the operating point closer to the critical region, which makes fractionation of methane from ethane difficult. For this reason, most demethanizers typically operate between 350 and 450 psig, with 500 psig as the upper limit.

To overcome the fractionation difficulty, Fluor has developed the twin-column high absorption process (TCHAP) using a dual-column approach. The first column operates as an absorber at 600 psig or higher pressure and is designed for bulk absorption. The second column, which serves as a demethanizer or deethanizer, operates at a lower pressure at about 450 psig. To improve NGL recovery, the overhead vapor from the second column is recycled using a small overhead compressor. The recycle gas is chilled using the absorber overhead vapor and used as a cold reflux to the absorber.

The process schematic of the TCHAP process is shown in Fig. 1.33. This NGL process may eliminate residue gas compression, which can

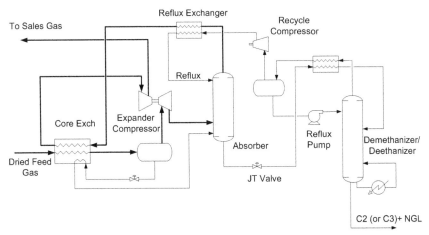

Figure 1.33 Fluor twin-column high absorption process (Mak et al., 2003). *J-T*, Joule—Thomson; *NGL*, natural gas liquid.

significantly reduce the overall power consumption of the facility. The TCHAP process can achieve over 80% ethane and 99% propane recovery (Mokhatab et al., 2015).

1.3.2.4.3.6 Fluor Twin-Reflux Absorption Process

NGL recovery units are frequently required to operate in ethane rejection mode when the profit margin of the ethane product is low. During these periods, the NGL recovery units are required to reject their ethane content to the sales gas pipeline. Ethane recovery processes, such as the GSP, can typically recover about 70–80% of the ethane content and about 98% of propane content. When operated in ethane rejection mode, some propane is lost with the rejected ethane, resulting in a loss of liquid revenue.

To circumvent this problem, Fluor has developed the twin-reflux absorption process (TRAP). The process can be operated on ethane recovery and can also operate in ethane rejection while maintaining high propane recovery (Mak et al., 2004a). The process flow schematic is shown in Fig. 1.34. The process is a dual-column design with the second column acting as a demethanizer during ethane recovery and as a deethanizer during ethane rejection. The front section of the process is similar to the GSP process. The feed gas is chilled using the cold residue gas and propane refrigeration and separated. About 30% of the expander drum vapor is subcooled forming a reflux stream to the absorber. The remaining expander drum vapor is expanded across a turboexpander producing the refrigeration.

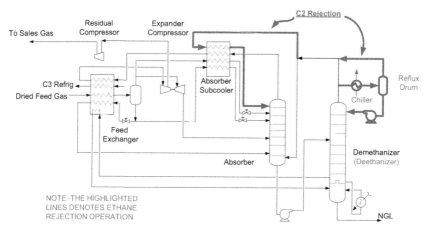

Figure 1.34 Fluor twin-reflux absorption process (Mak et al., 2004a). *NGL*, natural gas liquid.

The absorber produces a bottom liquid enriched in the C_{2+} liquids, which are routed to the second column. During ethane recovery, the second column operates as a demethanizer. The demethanizer overhead vapor is routed to the bottom of the absorber where the ethane content is reabsorbed. During this operation, the second column reflux system can be bypassed.

During ethane rejection, the second column operates as a deethanizer. The overhead vapor is rerouted from the absorber bottom to a deethanizer reflux system. The propane content is recovered and refluxed back to the deethanizer. The deethanizer reflux drum vapor is cooled in the absorber subcooler and recycled as a cold reflux to the absorber. During ethane rejection, the TRAP process can recover over 99% propane recovery (Mokhatab et al., 2015).

1.3.2.4.4 Other Hydrocarbons Removal Processes

For removal of small quantities of heavy hydrocarbons, such as in fuel gas conditioning, there are other simpler processes that can be used, instead of the traditional NGL recovery processes. Some of these processes are briefly described in the following sections.

1.3.2.4.4.1 **Solid Bed Adsorption** The solid bed adsorption process can be designed to selectively remove specific hydrocarbons. The adsorbent can be silica gel (i.e., Sorbead) that can be designed to remove most of the C_{6+} hydrocarbons. A typical two-bed adsorption process is shown in Fig. 1.35. Regeneration is accomplished by passing heated recycle gas

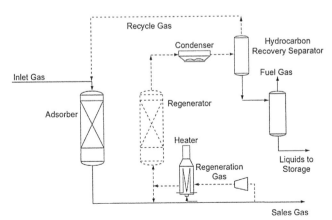

Figure 1.35 Schematic of a solid bed adsorption process for hydrocarbon removal (Mokhatab et al., 2015).

through the bed. The heavy hydrocarbon is recovered from the regeneration gas by cooling, condensation, and separation.

The solid bed adsorption process can be used to adsorb hydrocarbons at high pressure. This process has an advantage over the refrigeration process, which depends on low pressure for phase separation (Parsons and Templeman, 1990). The disadvantage is the HP vessels, which can be expensive. Details of the design and operation of a solid bed adsorption system can be found in Mokhatab et al. (2015).

1.3.2.4.4.2 Membrane Separation The membrane-based process is a compact design, particularly suitable for offshore installation. The process is simple and is suitable for unmanned operation. Fig. 1.36 shows the flow scheme of a membrane system used to produce a fuel gas for the engine of the pipeline compressor.

In this process, a slip stream of the pipeline gas is processed in the membrane module, which removes the heavy hydrocarbons, producing a lean gas as gas engine fuel. The hydrocarbons can be recycled to the compressor suction and recompressed into the pipeline system. Alternatively, the hydrocarbon contents can be recovered by condensation and chilling to produce a liquid by-product.

The hydrocarbon permeation membranes are typically made with vitreous polymers that exhibit high permeability with respect to heavy hydrocarbons (C_{6+}). New membrane materials are continuously being developed. Membrane Technology & Research Inc. has developed a new membrane that it claims to be applicable for separating C_{3+} hydrocarbons. However, the commercial application has yet to be proved.

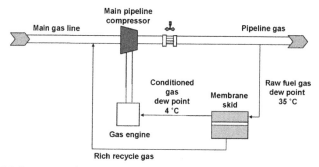

Figure 1.36 Process schematic of membrane separation process (Mokhatab et al., 2015).

1.3.2.4.4.3 Twister Supersonic Separation Twister separation technology is based on a supersonic mechanism (a combination of aerodynamics, thermodynamics, and fluid dynamics) to condense and remove water and heavy hydrocarbons from natural gas. It is based on the concept that condensation and separation at supersonic velocity reduces the residence time to milliseconds, allowing no time to form hydrates.

Twister supersonic separation technology can potentially offer significant costs and environmental benefits for offshore operation. More detailed discussion of the Twister technology can be found in Mokhatab et al. (2015).

1.3.2.5 NGL Fractionation

Once NGLs have been removed from the natural gas stream, they must be fractionated into their base components, which can be sold as high-purity products. Fractionation of the NGLs may take place in the gas plant but may also be performed downstream, usually in a regional NGL fractionation center. A typical process flow schematic is shown in Fig. 1.37. NGLs are fractionated by heating the mixed NGL stream and processing them through a series of distillation towers. Fractionation takes advantage of the differing boiling points of the various NGL components. As the NGL stream is heated, the lightest (lowest boiling point) NGL component boils off first and separates. The overhead vapor is condensed, a portion is used as reflux, and the remaining portion is routed to product storage. The heavier liquid mixture at the bottom of the first tower is routed to the second tower where the process is repeated and a different NGL component is separated

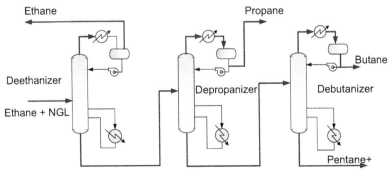

Figure 1.37 Typical natural gas liquid fractionation process (Mokhatab et al., 2015). *NGL*, natural gas liquid.

as product. This process is repeated until the NGLs have been separated into their individual components.

1.3.2.5.1 Fractionation Column Design and Operation

Fluctuation in feed conditions and compositions, which may happen on a daily basis, has significant impacts on the NGL fractionation unit operation. Their impacts on the design can be evaluated on the process simulators and the optimum operating variables can be assessed. The simulation software can now be integrated to the distributed control system (DCS) allowing real-time optimization.

Three books on distillation published by Kister (1989, 1992, 2006) can be used as references and guidelines in the design and operation of distillation columns. These books provide design parameters for equipment, discuss limitations of the design methods, and suggest solutions to troubleshoot columns. They also contain process calculations on column hydraulic and tower performance, tray and packing design details, and the methods to maximize column operating efficiency.

1.3.2.5.2 Liquid Products Processing

The liquid products from the NGL fractionation unit seldom meet customer's specifications without further treatment. The following sections discuss some of the processing methods to produce marketable liquid products.

1.3.2.5.2.1 NGL Contaminants Treating

If acidic and sulfur compounds are present in the feed gas and have not been removed before NGL recovery, then they will end up in the NGL products. The distribution of the contaminants for the various NGL products is summarized in Table 1.4. CO_2 and H_2S will show up in the deethanizer and depropanizer overheads, whereas most COS is concentrated in the propane fraction. The mercaptans tend to split between the various NGL fractions, depending on their molecular weight, with most of the high-molecular-weight mercaptans ending up in the condensate.

These contaminants can not only lead to odor problems but also form sulfur oxides on combustion. They can cause corrosion of equipment unless they are dehydrated. The presence of significant quantities of CO_2 can increase vapor pressure and lower heating value of the hydrocarbon liquids. COS and carbon disulfide (CS_2), although not corrosive in LPG, will hydrolyze slowly to H_2S, resulting in off-spec products.

Table 1.4 NGL contaminants distribution

	Ethane	Propane	Butane	Condensate
H_2O	x	x	x	x
CO_2	x	x		
H_2S	x	x		
COS	x	x		
CH_3SH	x	x	x	
C_2H_5SH	x	x	x	x
C_3H_7SH+		x	x	x
CS_2			x	x
Dimethyl sulfide			x	x
Others				x

Modified after John, M., Campbell & Co., September 2003.

1.3.2.5.2.1.1 Caustic Processes The hydrocarbon processing industry has historically used caustic solutions to extract or treat acidic impurities in liquid hydrocarbon streams. A number of caustic processes, both regenerative and nonregenerative, can be used to remove sulfur compounds from hydrocarbon liquids. The simplest process is the use of a nonregenerative solid potassium hydroxide (KOH) bed, which is effective for removal of H_2S but not for other sulfur compounds. One of the common processes for treating hydrocarbon liquids is the use of regenerative caustic wash with sodium hydroxide (NaOH). Typically the caustic wash is located downstream of an amine unit, which is used for bulk acid gas removal. Such arrangement would minimize the treating duty required by caustic wash. When acid gas quantities are small, a simple caustic wash is effective and economical. However, as the quantity of contaminants increases, the caustic wash process can be expensive due to high chemical and disposal costs.

Several organic materials may be added to caustic solutions (as promoters) to increase their solubility for mercaptans. Details on various promoter processes for caustic treating can be found in a paper by Maddox (1982). Mercaptans can be converted to disulfides by several methods. These disulfides will remain in the sweetened hydrocarbon product. The overall sulfur content, therefore, remains unchanged. However, the sulfur leaves as disulfide (no odor) rather than mercaptans. The method or combination of methods that can be used depends on the mercaptan content of the product to be treated and the specification that must be met (Fischer et al., 1993). Among the processes that convert (oxidize) mercaptans to disulfides, the "Doctor sweetening" process is the earliest. Doctor treatment ordinarily will leave 0.0004% mercaptan at which level there is negligible effect on the tetraethyl lead susceptibility of the gasoline. In this process, an alkaline solution of lead oxide (usually sodium plumbite) contacts the hydrocarbon stream forming lead mercaptides (soluble in the oil) with the mercaptans. The mixture is then treated with powdered sulfur (which has a high affinity for lead) and a conversion of the mercaptide into a so-called disulfide (which remains in solution in the gasoline stream) occurs. The reactions of the "Doctor treating" process are considered to be:

$$2RSH + Na_2PbO_2 \rightarrow (RS)_2Pb + 2NaOH \tag{1.1}$$

$$(RS)_2Pb + S \rightarrow R_2S_2 + PbS \tag{1.2}$$

Sulfur should be added in stoichiometric excess to achieve maximum conversion of the mercaptides to disulfides, where too much excess sulfur will cause formation of polysulfides (Maddox, 1982). Sometimes with no sulfur added, but in the presence of atmospheric oxygen and sodium hydroxide solution, the same conversion (Eq. 1.2) occurs, but only slowly, and not completely (McBryde, 1991).

The Merox process, a UOP trademark, developed and commercialized over 50 years ago by UOP, is used to treat end-product streams by rendering any mercaptan sulfur compounds inactive. This process can be used for treating LPG, gasoline, and heavier fractions. The method of treatment is the extraction reaction of the sour feedstock containing mercaptans (RSH) with caustic soda (NaOH) in a single, multistage extraction column using high efficiency trays (UOP's Extractor Plus). The extraction reaction is shown by the following equation:

$$RSH + NaOH \leftrightarrow NaSR + H_2O \tag{1.3}$$

After extraction, the extracted mercaptans in the form of sodium mercaptides (NaSR) are catalytically oxidized (by Merox WS catalyst, which is dispersed in the aqueous caustic solution) to water-insoluble disulfide oils (RSSR), as shown in the equation given below:

$$4NaSR + O_2 + 2H_2O \rightarrow 2RSSR + 4NaOH \tag{1.4}$$

The disulfide oil is decanted and sent to fuel or to further processing in a hydrotreater. The regenerated (lean) caustic is then recirculated to the extraction column. Typical product mercaptan levels can be controlled to less than 10 ppmw (UOP, 2003). A typical flow schematic of the UOP's Merox process is shown in Fig. 1.38.

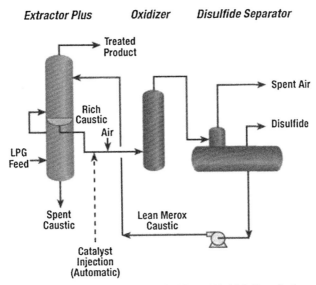

Figure 1.38 Liquid Merox treating process (UOP, 2003). *LPG,* liquefied petroleum gas.

The Sulfrex process (developed by Axens) is another technology similar to the Merox process, which has demonstrated its industrial performance, robustness, simple design, and ease in operation since 1975.

1.3.2.5.2.1.2 Molecular Sieve Technology Molecular sieve technology is commonly used for treating NGLs. Molecular sieves can be used for removal of sulfur compounds (H_2S, COS, and mercaptans) either in the gas or liquid phase. There are advantages and disadvantages for either option. The adsorber efficiency is higher in the gas phase since the mass transfer rate of sulfur compounds is much faster in the gas phase. However, the size and number of adsorbers are smaller for the liquid phase as the liquid rate tends to be significantly lower than the gas rate.

Molecular sieves can dry the feed stream in addition to contaminant removal. Their drawbacks are high capital and operating costs. The formation of COS is another concern if both H_2S and CO_2 are present. COS formation is to be avoided or minimized as much as possible as it converts back to H_2S in the presence of water, resulting in off-spec product and corrosion problems in downstream units. Different grades of molecular sieves have varying levels of conversion of H_2S to COS. Molecular sieve manufacturers can supply customized sieves to minimize COS formation. Usually these products are 3A-based, but they also have less adsorption capacity for CO_2 and H_2S.

For liquid phase adsorbers, adsorption flow tends to be upward, and regeneration downward. Molecular sieve regeneration is similar to that for gas phase application. In addition, liquid draining and filling time must be taken into account when determining the cycle times. If regeneration gas is supplied at a different pressure, pressurization and depressurization must be programmed in the setup. Care must be taken to ensure that maximum velocities are not exceeded, which may cause bed lifting (or fluidizing) and attrition of molecular sieves. More details on this process can be found in Mokhatab et al. (2015).

1.3.2.5.2.1.3 Amine Processes Amine treating is an attractive alternative, especially when an amine gas treating unit is already onsite. In such cases, the liquid-treating unit can often be operated using a slipstream of lean amine from the main amine regeneration unit.

Amine treating is often used upstream of caustic treaters to minimize caustic consumption caused by irreversible reactions with CO_2. In this process, H_2S and CO_2 from the sour liquid feed are absorbed by liquid–liquid contacting the sour liquid with the lean amine solvent. The amine unit design should consider the suitable amine solution and the type of contacting device. In most cases, generic amines (including DGA and DIPA, and the MDEA-based specialty solvents) will perform satisfactorily. The liquid–liquid contacting devices include packed towers, trayed towers, jet eductor mixers and static mixers. Most installations use random-packed columns as they are less expensive and avoid the potential back-mixing problems with tray column.

A typical LPG treating unit utilizing amine solvent is shown in Fig. 1.39. As can be seen, raw LPG enters the bottom of a packed absorber and lean amine enters the top of the absorber. Treated LPG leaves the absorber from the top and rich amine leaves the absorber from the bottom. The treated LPG is washed using a recirculating water wash system to recover the entrained amine and protect downstream caustic treaters. The treated LPG and the wash water are mixed in the water wash static mixer, which is then coalesced into two liquid phases and separated in the water wash separator. Makeup water is added to the circulating water wash circuit to maintain the concentration of the amine system.

COS is a stable, unreactive compound that is very difficult to reduce to concentration levels below 1 ppmv using a conventional amine and molecular sieve processes. The Amine-Di-Iso-Propanol process is a regenerative amine process to selectively reduce COS to very low levels (5 ppm wt as S) in liquid hydrocarbons such as LPG and NGL. Numerous

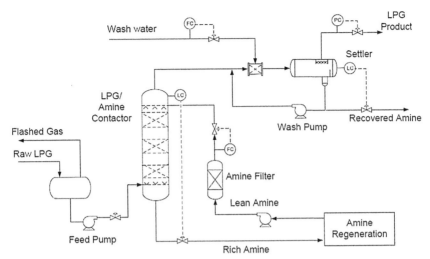

Figure 1.39 Process schematic for liquefied petroleum gas treating with amine solvent (Mokhatab et al., 2015). *LC*, level controller; *LPG*, liquefied petroleum gas.

nonregenerative metallic oxide processes are also available to remove COS from liquid products. Some of these processes remove the COS directly, and others require water to hydrolyze the COS to H_2S before it is reacted (Maddox, 1982).

1.3.2.5.2.2 Dehydration NGLs must be dehydrated to meet requirements of a handling chain to a direct consumer. The acceptable water content in light hydrocarbon liquid streams varies from no free water present to very low levels of moisture in liquid products. For example, most liquid sales specifications require the LPG to yield a negative result to the cobalt bromide test, which is equivalent to a water content of $15-30$ ppm.

Alumina is a satisfactory desiccant for liquid drying. If simultaneous treating and drying is required, molecular sieve type 13X is also a good candidate. There are other advanced processes to be used for contaminants removal and dehydration, which have been discussed by Mokhatab et al. (2015).

1.3.3 Integrating NGL Recovery and LNG Production Plants

NGL recovery plants can be operated with or without an LNG plant downstream. There are advantages of the stand-alone NGL recovery plants. They can be operated on NGL recovery, producing sales gas to the pipelines, and sending only the required volume to the LNG plants. On the other hand, the liquefaction plant can be designed to have the minimum gas

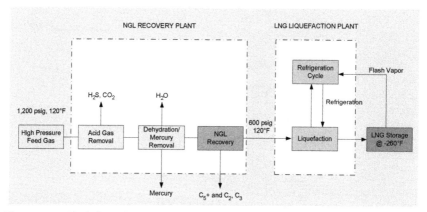

Figure 1.40 Block flow diagram of stand-alone natural gas liquid recovery and lique-faction plants (Mak et al., 2007). *LNG,* liquefied natural gas; *NGL,* natural gas liquid.

conditioning equipment that allows LNG production without the upstream NGL recovery plants. This independent operation increases the reliability and availability of both facilities but requires additional costs (Mokhatab et al., 2014). The process block flow diagram of a nonintegrated NGL recovery plant and an LNG liquefaction plant is shown in Fig. 1.40.

When compared with the stand-alone process, an integrated NGL/ LNG process can eliminate duplication of the heat exchangers and reduce the pressure drop between the two units (Mokhatab et al., 2014). Traditionally, the LNG plant is integrated with a scrub column for removal of the heavy hydrocarbons to avoid freeze-up in the cryogenic exchanger. A typical flow schematic is shown in Fig. 1.41. The scrub column normally operates at the feed gas pressure, typically at 700 psig. Because of the high pressure, the scrub column can recover only about 50—70% of the propane while producing a bottom product of the heavier components containing significant amounts of ethane and lighter components. Consequently, a deethanizer is required in addition to the traditional depropanizer and debutanizer in an NGL recovery plant. The deethanizer overhead containing the methane and ethane components are typically used in the fuel gas system. The scrub column uses propane refrigeration to generate reflux at about −20°F producing a lean gas that is sent to the main cryogenic heat exchanger.

The scrub column is simple and easy to operate; however, component separation is not very sharp and LPG recovery is limited due to the high operating pressure and relatively high reflux temperature. The advantage of a

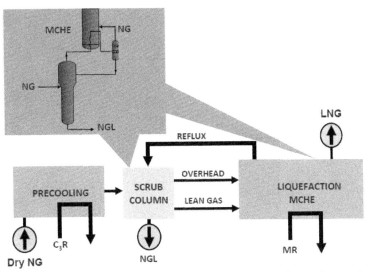

Figure 1.41 Integrated liquefied natural gas plant—scrub column (Hagyard et al., 2004). *LNG*, liquefied natural gas; *MCHE*, main cryogenic heat exchanger; *MR*, mixed refrigerant; *NG*, natural gas; *NGL*, natural gas liquid.

scrub column is that it can produce a HP residue gas to reduce liquefaction plant horsepower. On the other hand, the rough separation may be problematic with rich feed gas containing high levels of benzene and aromatics and heavier hydrocarbons. If they cannot be completely removed to the parts-per-million level, they may cause waxing and freeze-out in the cryogenic exchanger. For this reason, high propane recovery processes are necessary for LNG production to ensure complete removal of these heavier components, especially when processing rich gases.

There are many advanced NGL recovery processes that can be integrated with the liquefaction plant. However, these high propane recovery turboexpander processes typically use a demethanizer operating at 450–500 psig for good separation. Although these processes can recover 99% propane, they also require additional recompression to meet the LNG plant pressure requirement (Mokhatab et al., 2014).

If feed gas inlet pressure is high, say over 1200 psig, the turboexpander will need to operate at a higher expansion ratio, which also means that excessive refrigeration is generated that may not be fully utilized in the NGL recovery process. For this reason, a more efficient patented process has been developed by Fluor that uses a HP absorber operating at 600 psig or higher followed by a low-pressure deethanizer, with a recycle gas

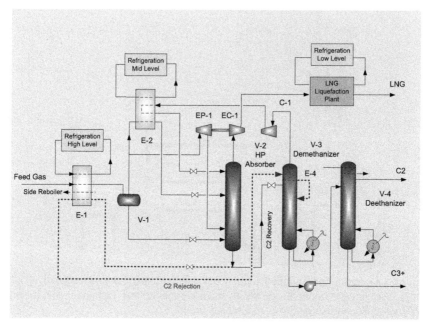

Figure 1.42 Fluor twin-column high absorption process natural gas liquid/liquefied natural gas integration (Mak et al., 2007). *HP*, high pressure; *LNG*, liquefied natural gas.

compressor integrated with the process. The Fluor TCHAP process can be used in the integrated plant and can achieve 99% propane recovery while producing a HP residue gas to the LNG plant. The Fluor TCHAP integrated process (Fig. 1.42) can save about 9% of the energy consumption for LNG production (Mak et al., 2007).

1.4 GAS PLANT SUPPORT SYSTEMS

1.4.1 Utility and Off-site

Operating the gas plants requires support from the utility and off-site systems. These typically include power generation, steam generation, heat medium supply, cooling water, instrument and plant air, nitrogen inert gas, fuel gas, relief, flare, potable water, and waste water treatment systems.

Small gas plants typically purchase electrical power off the local power grid. Larger gas plants or plants in remote areas generate their own power. For uninterrupted supply, power can be supplied from the power grid as a backup or using an emergency power generator. A diesel generator is typically required for plant black start. A battery power supply is required for life support during emergency.

Cogeneration plants are energy efficient and can be attractive options for reduction of operating costs, especially where the gas turbine exhaust can be used for heating or steam generation. Most large compressors in gas plants use gas engines or gas turbine drivers, which can be integrated with the steam system. Steam is used in gas plants for solvent regeneration and reboilers in NGL recovery and fractionation units. These units require cooling, and where available, cooling water is the most economical. In desert areas where water supply is not available, air coolers are the only choice of cooling.

A demineralizer unit is required to supply boiler feed-water makeup to steam generators and makeup to amine treating units. Wastewater must be treated in a wastewater treater before it can be discharged from the facility.

A reliable source of instrument air is critical to the gas plant operation. Typically, instrument air is supplied at about $100-120$ psig, and must be dried to meet the $-40°F$ water dew point. The air must be clean and dry to avoid water condensation and corrosion in the instrument air piping. One or multiple backup air compressors are required to ensure plant safety and reliability. Plant air can also be supplied from the same air compressors. Air receivers are sized to dampen out fluctuation in demand and to provide uninterrupted supply of instrument air.

Nitrogen is required for plant purging during start-up. It is also used for seal gas and for blanketing of storage tanks. Nitrogen consumption is typically very low in gas plant operation. Nitrogen can be produced by membrane separators or purchased from outside suppliers. In small gas plants, nitrogen can be supplied by nitrogen bottles.

High-pressure and low-pressure fuel gas are required to meet the fuel gas demands by steam boilers, hot-oil furnaces, and gas turbine drivers. Fuel gas must be treated and conditioned to meet the sulfur and heating value specifications.

1.4.2 Process Control Systems

Process control has always played a role in gas plants but has become more important over the years as companies try to reduce labor costs. Most plants use the DCS for individual units to provide both process control and operation history. Advanced process control (APC) systems, which are connected to the DCS, provide sophisticated plant control. The APC system uses multivariable algorithms that are programmed to perform simple online optimization. Another commonly used process control is the SCADA system. One important use of SCADA is the monitoring of field operations, with the capability of controlling the process unit equipment, flow valves, and compressor stations from the gas plant. Automation requires accurate input data to make the proper control decisions. These temperature, pressure, and flow sensors, and gas analyzers require constant maintenance and tuning to ensure accurate data are transmitted to the computer system.

Smaller plants may use programmable logic controllers.

1.4.3 Safety Systems

A relief and flare system must be provided to ensure plant safety. Multiple flares are typically installed to allow one flare to be maintained. The flare must be designed to protect against feed gas block discharge, as well as handling relief loads from power, and cooling water and other failure modes. Proper sizing of relief valves and rupture disks is critical to ensure that the process unit is protected during emergency. During normal operations, the gas plant is designed to avoid venting and flaring. If the plant is flaring mostly methane, the flame is bright but smokeless. If heavier hydrocarbons are flared, the flame will smoke and is noticeable. Smoke creates environmental problems. Steam can be injected to the flare system to maintain a smokeless flare. If the fuel has a low British thermal unit content, fuel gas can be added to ensure complete combustion.

1.5 OPTIMAL DESIGN AND OPERATIONS OF NATURAL GAS PROCESSING PLANTS

Efficient design and safe operation of the gas processing plant will increase the value of natural gas and hydrocarbon liquid products. To achieve this, the complete process design should be carried out in three stages: process modeling and simulation, process control, and process optimization.

1.5.1 Process Modeling and Simulation

Process modeling, which involves the development of a set of equations that help gain insight to the behavior of a given process or unit operation, is one of the principal steps in process design. Modeling is used in the natural gas industry for simulations of process and equipment design, analysis of system behavior, operator training, leak detection, and design of controllers.

There are two main classifications of mathematical models: steady-state models and dynamic models. In steady-state models, which are commonly used to investigate process behavior during normal operation, individual process units or entire flow sheets are modeled such that there are no time deviations of variables and parameters. Dynamic modeling, which accounts for process transients from an initial state to a final state, is primarily used in the study of process start-up, shutdown, and controller design. Dynamic models can also be used for steady-state analyses in certain special cases, where there is a need to evaluate only a few steady-state cases. This is usually beneficial if a dynamic model is already available and can be leveraged for the intended purpose.

A clear distinction needs to be made between models and simulation. Models are tools that are used for various purposes, simulation being one of them. Simulation simply means solving a set of process modeling equations that define a process or unit operation to obtain a solution for unknown process variables (Aneke, 2012). Both steady-state and dynamic simulations are carried out in the various stages of the process life cycle. These simulations are useful for a variety of purposes, including but not limited to engineering and process studies, control system studies, and applications in day-to-day operations (Broussard, 2002). Dynamic simulation has established itself as a valuable technology in the natural gas processing industry. The operability and profitability of the gas plant during its life depends on good process and control system design. Dynamic simulation helps to ensure that these aspects are analyzed early in the plant design stage. This helps to eliminate any costly rework that may be needed later. The various engineering contractors have used dynamic simulation extensively for improving the plant design (Valappil et al., 2005; Patel et al., 2007; Masuda et al., 2012). Also, process licensors have used these tools to improve and optimize their offerings. Apart from these, the gas plant owner-operators have also found good use of dynamic simulation, especially in the form of operator training simulators.

The details of developing gas plant models, including the modeling of individual unit operations, and their applications in design and operations of plants will be explained in the subsequent chapters.

1.5.2 Process Control

Processing natural gas to profitably produce end products requires precision and is potentially hazardous. Small changes in a process can have a large impact on the end result. Conditions such as pressure, temperature, flow, composition, and many other factors must be controlled carefully and consistently within specified limits to ensure quality and safety. A well-designed, maintained, and leveraged control system will reduce start-up time, maintain maximum operating profit, avoid forced shutdowns, keep operating and maintenance costs as low as possible, assist in the management of environmental compliance, and support plant safety and security needs without adversely affecting construction costs.

An overview of the main control functionalities in the gas processing plant and their applications are described in subsequent chapters.

1.5.3 Process Optimization

Variations in process operating conditions give rise to variations in process performance. This is the problem that process optimization seeks to solve. Process optimization is therefore an iterative process simulation to find the best solution to a given process within certain defined operating constraints (Aneke, 2012).

Gas processing operations constantly experience changing conditions due to varying contracts, feed rates, feed compositions, and pricing. To capture the maximum entitlement measured in profits, these operations are prime candidates for real-time optimization (Bullin and Hall, 2000). Real-time optimization enables operating facilities to respond efficiently and effectively to the constantly changing conditions of feed rates and composition, equipment condition, and dynamic processing economics. In fact, world-class gas processing operations have learned how to optimize in real time to return maximum value to their stakeholders. Applications of real-time optimization of gas processing facilities have been adopted. Advances in computer power, robust modeling approaches, and the availability of real-time pricing have enabled this technology. An online optimization model also provides a continuously current model for accurate simulations required for offline evaluations and studies. Equipment conditions including fouling factors for heat exchangers and deviation from efficiencies predicted by head curves for rotating equipment are tracked over time. The impact of additional streams under different contractual terms can be evaluated with the most up-to-date process model available.

The concepts of real-time optimization as well as the considerations for successful application in the gas processing industry will be described in the subsequent chapters.

REFERENCES

Abbott, J., Oppenshaw, P., March 11–13, 2002. Mercury removal technology and its applications. In: Paper Presented at the 81st Annual GPA Convention, Dallas, TX, USA.

Aneke, M.C., 2012. Process modeling and simulation. In: Holloway, M.D., Nwaoha, C., Onyewuenyi, O.A. (Eds.), Process Plant Equipment: Operation, Control, and Reliability, first ed. John Wiley & Sons, Inc., New Jersey, NJ, USA.

Bradley, A., De Oude, M., van der Zwet, G., August 23–27, 2009. Sulfinol-x process-efficiently achieving complex contaminant removal objectives. In: Paper Presented at the 8th World Congress of Chemical Engineering (WCCE8), Montreal, Canada.

Broussard, M., May 2002. Maximizing simulation's reach. Chemical Engineering 50–55.

Bucklin, R., December 20, 1966. Method and Equipment for Treating Hydrocarbon Gases for Pressure Reduction and Condensate Recovery. U.S Patent No. 3292380.

Bullin, K.A., Hall, K.R., March 13–15, 2000. Optimization of natural gas processing plants including business aspects. In: Paper Presented at the 79th GPA Annual Convention, Atlanta, GA, USA.

Carlsson, A.F., Last, T., Smit, C.J., March 11–14, 2007. Design and operation of sour gas treating plants for H_2S, CO_2, COS and mercaptans. In: Paper Presented at the 86th Annual GPA Convention, San Antonio, TX, USA.

Carnell, P.J.H., Row, V.A., April 24–27, 2007. A rethink of the mercury removal problem for LNG plants. In: Poster Presented at the 15th International Conference & Exhibition on Liquefied Natural Gas (LNG 15), Barcelona, Spain.

Clarke, D.S., Sibal, P.W., March 16–18, 1998. Gas treating alternatives for LNG plants. In: Paper Presented at the 77th Annual GPA Convention, Dallas, TX, USA.

Eckersley, N., 2010. Advanced mercury removal technologies. Hydrocarbon Processing 89 (1), 29–35.

Edmonds, B., Moorwood, R.A.S., Szcepanski, R., March 1996. Mercury partitioning in natural gases and condensates. In: Paper Presented at the GPA Europe Conference, London, UK.

Fischer, E., Goel, R., Saunders, D., March 15–17, 1993. Preliminary process selection for natural gas liquid (NGL) treating. In: Paper Presented at the 72nd GPA Annual Convention, San Antonio, TX, USA.

GPSA Engineering Data Book, 2004. Gas Processors Suppliers Association (GPSA), twelfth ed. Tulsa, OK, USA.

Grootjans, H., November 21–23, 2005. Sustained development in the Pearl GTL treating design. In: Paper Presented at the 2005 International Petroleum Technology Conference, Doha, Qatar.

Hagyard, P., Paradowski, H., Gadelle, D., Morin, P., Garcel, J.C., March 21–24, 2004. Simultaneous production of LNG and NGL. In: Paper Presented at the 14th International Conference & Exhibition on Liquefied Natural Gas (LNG 14), Doha, Qatar.

Harvey, C.G., Verloop, J., 1976. Experience confirms adaptability of the SCOT process. In: Paper Presented at the 2nd International Conference of the European Federation of Chemical Engineers, University of Salford, Manchester, UK.

Hubbard, R., August 2009. The role of gas processing in the natural-gas value chain. Journal of Petroleum Technology 65–71.

Hudson, C., April 18–21, 2010. Implications of mercury removal bed material changeout: brownfield versus greenfield. In: Poster Presented at the 16th International Conference & Exhibition on Liquefied Natural Gas (LNG 16), Oran, Algeria.

Iyengar, J.N., Johnson, J.E., O'Neill, M.V., March 13–15, 1995. Sulfur emissions identification and handling for today's SRU/TGU facilities. In: Paper Presented at the 74th Annual GPA Convention, San Antonio, TX, USA.

John, M., Campbell & Co, September 2003. Technical assistance service for the design, operation, and maintenance of gas plants. In: Course Materials Presented at the BP Exploration Company Columbia Ltd., Columbia.

Johnson, J.E., November 26—29, 2005. Hazards of molten and solid sulfur storage, forming, and handling. In: Paper Presented at the 2005 Sour Oil & Gas Advanced Technology Conference, Abu Dhabi, UAE.

Katz, D.L., Cornell, R., Kobayashi, R., Poettmann, F.H., Vary, J.A., Elenblass, J.R., Weinaug, C.G., 1959. Handbook of Natural Gas Engineering. McGraw-Hill Book Company, New York, NY, USA.

Kidnay, A.J., Parrish, W.R., 2006. Fundamentals of Natural Gas Processing, first ed. CRC Press, Boca Raton, FL, USA.

Kister, H.Z., 1989. Distillation Operation. McGraw-Hill, New York, NY, USA.

Kister, H.Z., 1992. Distillation Design. McGraw-Hill, New York, NY, USA.

Kister, H.Z., 2006. Distillation Troubleshooting. John Wiley & Sons Inc., Hoboken, NJ, USA.

Klinkenbijl, J.M., Dillon, M.L., Heyman, E.C., March 1—3, 1999. Gas pre-treatment and their impact on liquefaction processes. In: Paper Presented at the 78th Annual GPA Convention, Nashville, TN, USA.

Kohl, A., Nielsen, R., 1997. Gas Purification, fifth ed. Gulf Publishing Company, Houston, TX, USA.

MacKenzie, D., Cheta, I., Burns, D., November 2002. Removing nitrogen. Hydrocarbon Engineering 7, 57—63.

Maddox, R.N., 1982. Gas and Liquid Sweetening, third ed. Campbell Petroleum Series, Norman, OK, USA.

Mak, J.Y., Deng, E., Nielsen, R.B., August 5, 2003. Methods and Apparatus for High Propane Recovery. U.S. Patent No. 6601406.

Mak, J.Y., Chung, W., Graham, C., Wierenga, D., March 14—17, 2004a. Retrofit of a NGL process for high propane and ethane recovery. In: Paper Presented at the 83rd GPA Annual Convention, New Orleans, LA, USA.

Mak, J.Y., Nielsen, D., Graham, C., Schulte, D., March 14—17, 2004b. A new and flexible LNG regasification plant. In: Paper Presented at the 83rd GPA Annual Convention, New Orleans, LA, USA.

Mak, J.Y., January 4, 2005. High Propane Recovery Process and Configurations. U.S. Patent No. 6837070.

Mak, J.Y., Nielsen, R.B., Graham, C., February 22—23, 2007. A new integrated NGL recovery/LNG liquefaction process. In: Paper Presented at the GPA Europe Conference, Paris, France.

Mak, J.Y., Nielsen, R.B., Chow, T.K., September/October 2010. Zero Claus plant SOx emissions. Sulphur 331.

Mak, J.Y., Row, A.R., Varnado, C., April 15—18, 2012. Production of pipeline gas from a raw gas with a high and variable acid gas content. In: Paper Presented at the 91st Annual GPA Convention, New Orleans, LA, USA.

Markovs, J., Clark, K., March 13—16, 2005. Optimized mercury removal in gas plants. In: Paper Presented at the 84th Annual GPA Convention, San Antonio, TX, USA.

Masuda, K., Nakamura, M., Momose, T., October 8—11, 2012. The use of advanced dynamic simulation technology in the engineering of natural gas liquefaction plants. In: Paper Presented at the Gastech 2012 Conference & Exhibition, London, UK.

McBryde, W.A.E., 1991. Petroleum deodorized: early Canadian history of the 'doctor sweetening' process. Annals of Science 48, 103—111.

Millward, R.J., Finn, A.J., Kennett, A.J., September 21—23, 2005. Pakistan nitrogen removal plant increases gas quality. In: Paper Presented at the GPA Europe Annual Conference, Warsaw, Poland.

Mock, J.M., Hahn, P., Ramani, R., Messersmith, D., February 24—27, 2008. Experiences in the operation of dehydration and mercury removal systems in LNG trains. In: Paper Presented at the 58th Annual Laurance Reid Gas Conditioning Conference (LRGCC), Norman, OK, USA.

Mokhatab, S., Hawes, P., April 22—24, 2015. Optimal mercaptans removal solution in gas processing plants. In: Paper Presented at the GPA Europe Spring Technical Conference, Hamburg, Germany.

Mokhatab, S., Meyer, P., May 14—15, 2009. Selecting best technology lineup for designing gas processing units. In: Paper Presented at the GPA Europe Sour Gas Processing Conference, Sitges, Barcelona, Spain.

Mokhatab, S., Mak, J.Y., Valappil, J.V., Wood, D.A., 2014. Handbook of Liquefied Natural Gas. Gulf Professional Publishing, Burlington, MA, USA.

Mokhatab, S., Poe, W.A., Mak, J.Y., 2015. Handbook of Natural Gas Transmission and Processing, third ed. Gulf Professional Publishing, Burlington, MA, USA.

Northrop, S., Sundaram, N., August 2008. Modified cycles, adsorbents improve gas treatment, increase mole-sieve life. Oil & Gas Journal 106, 29.

Nougayrede, J., Voirin, R., July 17, 1989. Liquid catalyst efficiently removes H_2S from liquid sulfur. Oil & Gas Journal 65—69.

Parsons, P.J., Templeman, J.J., 1990. Models performance leads to adsorption-unit modifications. Oil & Gas Journal 88 (26), 40—44.

Patel, V., Feng, J., Dasgupta, S., Ramdoss, P., Wu, J., September 10—13, 2007. Application of dynamic simulation in the design, operation, and troubleshooting of compressor systems. In: Paper Presented at the 36th Turbomachinery Symposium, Houston, TX, USA.

Pitman, R.N., Hudson, H.M., Wilkinson, J.D., Cuellar, K.T., March 16—18, 1998. Next generation processes for NGL/LPG recovery. In: Paper Presented at the 77th GPA Annual Convention, Dallas, TX, USA.

Row, V.A., Humphrys, M., May 25—27, 2011. The impact of mercury on gas processing plant assets and its removal. In: Paper Presented at the GPA Europe Spring Conference, Copenhagen, Denmark.

Schulte, D., Graham, C., Nielsen, D., Almuhairi, A.H., Kassamali, N., March 31—April 01, 2009. The Shah Gas Development (SGD) Project — a new benchmark. In: Paper Presented at the 5th International Sour Oil & Gas Advanced Technology Conference, Abu Dhabi, UAE.

Shell DEP 31.40.10.12—Gen, 1998. Design manual of multiple-pipe slug catchers. In: Design and Engineering Practice (DEP) Publications, Shell Global Solutions International B.V., The Hague, The Netherlands.

Son, K.V., Wallace, C., November 7—9, 2000. Reclamation/regeneration of glycols used for hydrate inhibition. In: Paper Presented at the 12th Annual Deep Offshore Technology Conference, New Orleans, LA, USA.

Stone, J.B., Jones, G.N., Denton, R.D., March 11—13, 1996. Selection of an acid-gas removal process for an LNG plant. In: Paper Presented at the 75th Annual GPA Convention, Denver, CO, USA.

Thompson, G.R., Adams, J.B., Hammadi, A.A., Sibal, P.W., March 21—24, 2004. Qatargas II: full supply chain overview. In: Paper Presented at the 14th International Conference and Exhibition on Liquefied Natural Gas (LNG 14) Conference, Doha, Qatar.

Trautmann, S., Davis, R., Harris, C., Ayala, L., May 10—12, 2000. Cryogenic technology for nitrogen rejection from variable content natural gas. In: Paper Presented at the XIV Contencion Internacional de Gas, Caracas, Venezuela.

UOP, 2003. Merox process for mercaptan extraction. In: UOP 4223—3 Process Technology and Equipment Manual. UOP LLC, Des Plaines, IL, USA.

Valappil, J.V., Mehrotra, V., Messersmith, D., 2005. LNG lifecycle simulation. Hydrocarbon Engineering 10 (10), 27–34.

Wilkinson, D., Johnson, G., April 28, 2010. Nitrogen rejection technology for Abu Dhabi. In: Paper Presented at 18th Annual GPA Gulf-Chapter Conference, Muscat, Oman.

Wong, V., Mak, J.Y., Chow, T., March/April 2007. Fluor technology for lean acid gas treatment. Sulphur Magazine 309, 39–42.

CHAPTER 2

Process Modeling and Simulation

2.1 INTRODUCTION

2.1.1 Definition of Process Simulation

At the most abstract and simple level, a process simulation model is a representation of a chemical process plant to facilitate design, analysis, or other types of studies of the behavior of that plant. Often this refers to creating a mathematical representation using computer software. All simulation models share several basic characteristics:

- Representation of the thermodynamic behavior of material
- A representation of equipment that is encountered in a chemical process plant
- Representation of chemical reaction connections between equipment to represent the flow of material

The nature and the level of detail of each representation can vary greatly and the user of the simulation model must make a choice such that the results and accuracy of the model satisfy the needs of the task at hand. A model should be fit for purpose, although care should be taken not to overly simplify the model and in that way overlook possible issues in design or operation.

2.1.2 Benefits of Simulation in Gas Processing Plants

The benefits of simulation are not unique for gas processing plants. They apply to any chemical process plant. Simulation models bring value throughout the complete life cycle of a process. At the very early conceptual design phase of the process, the emphasis will be more on relatively simple heat and material balances. As the definition of the plant becomes more detailed, so will the model.

During the conceptual phase, the results of the simulation will be used to determine approximate equipment sizes, and power and utility consumption. This will enable an estimate of the investment required for the plant and of the operating costs. The combination of these numbers with

Modeling, Control, and Optimization of Natural Gas Processing Plants
ISBN 978-0-12-802961-9
http://dx.doi.org/10.1016/B978-0-12-802961-9.00002-4

the cost of raw material and the expected market prices of the products will determine if the project is economically viable.

During the front-end engineering design phase, the simulation model will provide sufficient information for a detailed design of each piece of equipment, the piping that will connect them, and the instrumentation. Dynamic simulation models will also provide insight into the controllability of the proposed design.

In the detailed engineering phase and through the start-up of the plant, dynamic simulation models will provide information on the tuning of the process control system and on the validity of the proposed start-up scenarios and may help the training of the engineers and operators on the operation of the plant. Another very important aspect is the use of models to define safety equipment and to ensure that the safety system is designed to protect the plant and the people operating the plant under all incident scenarios.

Finally, simulation models are invaluable during the operation of the plant for monitoring the condition of the equipment, trouble shooting, and optimizing the performance as a function of current feed product quality, market prices, and environmental factors.

2.1.3 Chapter Objectives

This chapter aims to provide guidance in selecting the model nature and level of detail such that it is fit for the purpose, and can also be reused if more demanding tasks arise in the future. It will also discuss modeling best practices. These practices are aimed at faster model development while developing robust models with the required accuracy.

2.2 THERMODYNAMICS

At the heart of any model is the description of the behavior of the fluids that are to be processed when subject to a wide range of temperatures and pressures. The accuracy of the chosen thermodynamic model is key to the accuracy of the complete model. The selection of a thermodynamic model will depend on the nature of the components to be represented and the range of temperatures and pressures that are to be considered. Hydrocarbon molecules behave in a relatively ideal fashion at low pressure and near-ambient temperature. However, most gas processing will involve high-pressure operations and sometimes cryogenic temperatures. To properly model the behavior at high pressure, an equation of state (EOS) should be used. The most popular EOSs for gas processing are Peng Robinson and Soave—Redlich—Kwong (SRK).

When choosing a model, the complexity of the model should be taken into account. As the thermodynamic routines are called virtually all the time in a simulation model, a complex thermodynamic model will slow down the model execution.

High-accuracy EOSs exist for hydrocarbon mixtures. The RefProp model from the National Institute of Standards and Technology is an example of such an EOS. This model is seldom used to model a complete flow sheet. The two reasons are that the model consumes much more central processing unit (CPU) power than an EOS like Peng Robinson and the model is less robust. The reduced robustness may lead to convergence issues when the operating conditions for which the calculations are performed fall outside of the normally expected conditions. Therefore, it is best to limit the use of complex thermodynamic models to parts of the simulation model that require that specific model to produce meaningful results. Care should be taken to avoid including these calculations in loops with many iterations.

For example, the mass density of the gas may be calculated more accurately with a complex EOS. However, the gain in accuracy will often only be a few percent. For the engineering design calculations this will not make a meaningful difference. But when determining the export rate of the gas, a 1% error in mass density means a 1% impact on revenue and this translates into a considerable amount of money. However, even in this case, it is only important to calculate an accurate mass density at the export point and not throughout the whole process.

In gas processing, some of the units that require special attention concerning thermodynamics are triethylene glycol (TEG) dehydration, gas sweetening, monoethylene glycol (MEG) regeneration, and cryogenic operations. The treatment of liquid water streams in general may also require special attention. More specific information can be found in the case studies section.

An important element that needs to be kept in mind when selecting a thermodynamic model is that the thermodynamic model itself is only a framework represented through a set of equations. All of these equations require component-specific parameters and often also parameters that describe the interaction between components. If those parameters are not known, even the best thermodynamic model is worthless!

2.3 STEADY-STATE VERSUS DYNAMIC MODELS

Once the thermodynamic model has been selected, the next choice to be made when embarking on a modeling project is whether the task at hand requires a steady-state or a dynamic simulation model. The difference between these two types of models is that a steady-state model assumes the plant is operating in a perfectly stable manner. Variation with respect to time is assumed to be zero. The vast majority of simulation models used are of the steady-state type. The reason for this popularity is that the assumption of a zero derivative with respect to time simplifies the modeling of all equipment enormously. This has a significant impact on model solution time. It also allows for a lot more flexibility in the specification of the model. In a steady-state model, it is very easy to define the model as a task to be accomplished in relatively abstract terms. For example, the process may require the pressure and temperature to be changed. A steady-state model can usually do this with a single unit operation. Although it is possible to create a similar abstract dynamic model, if the actual equipment function and geometry is not known, the dynamics of the model are arbitrary and may even lead to wrong conclusions.

A dynamic simulation is typically aimed at a more realistic and more detailed model of the behavior of the process and it includes how the process behaves over time. If we consider a simple vapor—liquid separation vessel, then in a steady-state simulation the nature of the model and the inlet and outlet connections are all that is required to define the simulation. The volume in the vessel or the liquid level has no impact on the results. In a dynamic simulation, the volume and liquid level are key elements of the simulation. The aim of the simulation could, for example, be to study the level control of the vessel or the pressure transients in the vessel. The additional effort required to create a dynamic simulation model is the second reason why steady-state models are used more than dynamic models.

For any modeling effort that is aimed at main equipment design, the initial modeling choice will always be a steady-state model. A dynamic model may be used at a later stage to validate the process behavior given the design choices resulting from the steady-state model combined with specific circumstances. If the aim of a simulation is related to process control or to safety, a dynamic simulation is used as the phenomena to be studied are always varying in time.

2.4 SIMULATION OBJECTIVES VERSUS MODELING EFFORT

2.4.1 Shortcut Versus Rigorous Models

Shortcut models tend to describe the process in more abstract terms; the unit operations used do not necessarily directly represent actual plant equipment but sometimes a combination of several equipment items. Rigorous models will model each equipment item individually.

There is no single definition of rigorous. Rigor can be introduced by increasing the number of unit operations used to model the system and it can also be introduced by changing the methods used inside of a unit operations model to represent the function of the equipment.

As an illustration, consider a small part of the plant that has a pump to increase the pressure of a liquid and a heat exchanger that reduces the temperature of the liquid. A shortcut model could represent this as a single unit operation. The model would provide limited information for the design of the pump or the heat exchanger, but it does bring the liquid to the expected state to feed it in the next unit operation. A first step toward a more rigorous model would be to model this as a pump and a cooler. A next step would be to use a heat exchanger model that also represents the cooling fluid. The model could be further detailed by including geometry information for the heat exchanger or pump performance curves.

For a dynamic model some the control valves should be added such that the behavior of controllers can be checked. The model detail can be extended even more by also including the spare pump that will be installed, the block valve that will be included, the minimum flow line on the pump, and the bypass circuit on the exchanger. Additional details like the inertia of the pump can be added for studying the start-up of the pump.

The choice of the level of detail used depends on the purpose of the model. Clearly, the shortcut model will not be of any help investigating the behavior of the system when it switches to the spare pump in case of failure of the normally operating pump. However, if the currently required information is pump power and utilities consumption, a model slightly more rigorous than the shortcut model is sufficient and additional modeling of the spares and bypasses would be a waste of effort.

Another typical example would be the case where a larger plant includes a TEG dehydration unit and it is already known that this will be subcontracted to a specialist vendor. The main model can represent the whole TEG dehydration unit as a single simplified block that produces a dehydrated gas with the water content specification demanded from the specialist vendor.

2.4.2 Lumped Parameter Versus Distributed Models

Lumped parameter and distributed models come into play when considering rigorous equipment models. It is a continuation of the sort of choices that are made when choosing between shortcut and rigorous models at the level of a single piece of equipment. A totally lumped parameter model will consider the content of one equipment to be homogeneous; it will not consider radial or axial gradients in the fluid properties. A fully distributed model will consider the variation of fluid properties and interaction with its neighboring elements in all three dimensions and over time. Computational fluid dynamics is an example of a method that can be a fully distributed model.

As with shortcut and rigorous process models, there is a continuum of variations between the two extremes. This can also best be illustrated with some examples. A totally lumped model could be a phase separation vessel model that only predicts phase separation and a uniform heat input from the environment. This would be a zero dimensional model.

A first (small) step toward a distributed model is to calculate the heat input for the part of the vessel that contains the vapor phase and the part that contains the liquid phase separately. The next step could be to drop the assumption of phase equilibrium and to calculate phase compositions and temperatures using the rate of exchange of material and heat between the phases. A practical example of this is the selection of the model used for determining the result of depressuring a vessel over time. If the calculated temperatures approach the limit where a different material of construction should be used, it would be prudent to increase the rigor of the model and hence to move to a more distributed model.

Next we could consider the problem of the behavior of a large liquefied natural gas tank. A known phenomenon is roll over. The model that may have been satisfactory to describe vessel depressuring would not capture the rollover phenomenon. The model would also consider that in the liquid phase there can be multiple layers with different compositions and properties. This would lead to a one-dimensional model. If the effect of the pitch and roll of the ship are lumped together in a single parameter, our previous one-dimesional model would be expanded to be a two-dimensional model. If the effects of pitch and roll are quantified separately, then the resulting model would be a three-dimensional model.

The type of model selected should always depend on the purpose of the model and ultimately it will be the economics that determine the choice. If it takes a 1 million dollar study to quantify the additional effect of the sun heating a vessel in a nonuniform way and the potential gain from the results of the study is 25,000 dollars per year, it is highly unlikely that such a study would be done.

2.4.3 Commercial Models Versus Bespoke Models

As a bespoke model will typically be more expensive than an off-the-shelf model, a bespoke model would only be selected if off-the-shelf models cannot provide the required functionality of the model. An intermediate solution should also be investigated. Rather than creating a model from scratch, it may be more economical to use an off-the-shelf tool and construct the required model from elements provided by the off-the-shelf tool. Another element may come into play here. If the new model will likely be useful for several applications, then the added cost may be offset by the cost one would otherwise incur by reimplementing a particular solution over and over again in an off-the-shelf tool. Another important factor may be that once the bespoke model has been validated, it does not require revalidation for the next applications.

2.5 PROCESS SIMULATION APPROACHES

For both steady-state and dynamic models the approach can be split into two categories: modular models and equation solver type models. As nothing is black and white in this world, there is also scope for approaches that are a hybrid between the two.

2.5.1 Modular Approach for Steady-State Models

A modular approach is by far the most used approach for steady-state process modeling. Most commercial simulators fall into this category or have this capability in the offering. Every piece of equipment or combination thereof is modeled as a separate block. Inside the block itself different approaches can be used, but the key distinguishing factor is that a block will only attempt to solve if sufficient information has been made available through user input and from upstream or downstream blocks. If a block can use information from both upstream and downstream blocks, the simulator is said to be nonsequential. This capability can be used to create computationally more efficient models.

An example application could be the model of a pump followed by a generic cooling device. With a sequential modular approach, the feed stream of the pump needs to be completely known. This typically means composition, pressure, temperature, and flow are known. The pump then needs some information that allows it to determine the discharge pressure and a pump efficiency to solve. The generic cooling device needs the results from the pump and information to determine the cooling duty and the outlet pressure.

Besides the feed definition, the user input could, for example, be pump discharge pressure, pump efficiency, cooler pressure drop, and cooler outlet temperature. If the user has as information the pump efficiency, cooler pressure drop, cooler outlet pressure, and outlet temperature, an iterative procedure is required to find the solution to this small problem. With a nonsequential modular solver, the cooler would be able to partially solve and determine the inlet pressure from the outlet pressure and the pressure drop. This would enable the pump to solve completely and subsequently the cooler would be able to solve for the remaining unknowns.

The modular approach has the advantage of being very robust. The robustness, however, comes at the price of speed of solution, in particular if the model involves recycle streams or includes some sort of iterative instruction that varies a parameter to reach a specified target or if the goal is to perform an optimization of the modeled plant.

2.5.2 Equation Solver Approach for Steady-State Models

In the equation solver approach for steady-state models all the calculations required for each of the blocks are expressed as equations. The connections between the blocks are also expressed as equations. The solution algorithm is a mathematical equation solver. One advantage of this approach is that the user input does not need to be such that at least part of a block can be solved to solve the complete flow sheet.

Going back to the example of the pump and the cooling device, even the nonsequential modular solver is not able to solve the problem if the information available to the user is the cooler outlet temperature, outlet pressure, and duty. The cooler can determine the inlet molar enthalpy, but the pump model is not set up to derive the discharge pressure from the molar enthalpy. An equation solver would not require such a specific setup. The equations that govern the pump behavior will either directly or indirectly link the pump outlet molar enthalpy to the pump outlet pressure and be able to solve for it. At the same time, this example illustrates the reduced robustness of this solution technique. As the molar enthalpy of the pump outlet is rather insensitive to the pressure, it is very easy to specify a combination of inputs that will be infeasible.

The big advantage of the equation solver approach is speed. The solution of problems with recycles or targets or optimizations does not take substantially longer than the solution of a flow sheet that does not have this type of added complexity. Because of this combination of advantages and disadvantages, the main area of use of this type of flow sheet solvers is in applications where optimization is a requirement. Online optimization applications are the main applications of this type of simulators.

2.5.3 Combined Approach in Steady-State Models

Although most commercial steady-state simulators are said to be modular, they frequently have some specific equation solver type solutions inside them. The main example found in virtually all steady-state simulators is the modeling of the distillation column. A distillation model consists of a stacked series of phase-separation stages that are highly interlinked. If this model were to be solved in a strictly modular fashion, a model with N separation stages would require (N-1) recycle or tear streams. The solution time would be very long and quite often no solution would be found. So, for this part of the model a specific equation solver is used. As the functionality is limited to certain types of equipment, the solver can be optimized to be not only fast, but also robust. Despite this, the modeling of a distillation column is often one of the more challenging parts of creating a flow sheet.

2.5.4 Modular Approach for Dynamic Models

A modular approach is not very often used for dynamic models, at least, not for gas processing applications. The main reason is that dynamic models very often also capture the hydraulic behavior of a plant. The hydraulics of a plant is determined by the combination of pressures, flows, holdups, and flow resistances of a series of equipment. A change of pressure at one end of the plant can affect the pressures and flows of large parts of the plant, and this happens quasi-instantaneously. A modular solution technique is not suitable at all for this type of problem.

In a pharmaceutical plant, the main dynamics of interest is the behavior of liquid holdups and reactors and a modular approach is appropriate here. In a gas processing plant, the hydraulic behavior of fluids is very important and hence no dynamic simulator for gas processing applications uses a purely modular approach.

2.5.5 Equation Solver Approach for Dynamic Models

The equation solver approach for dynamic simulation is comparable with the equation solver approach for steady-state simulators. However, the disadvantage in robustness is combined with a lesser performance when compared with the hybrid approach described later. The biggest CPU power consumer in any process simulation model, be it steady state or dynamic, are the flash and thermodynamic calculations. With an equation solver approach for dynamic modeling, these time-consuming calculations are performed simultaneously and at the same frequency of all other calculations. This usually results in a comparatively poor overall performance. The equation solver approach is used for dynamic modeling if high accuracy is the primary concern.

2.5.6 Hybrid Approach for Dynamic Models

The hybrid approach combines an equation solver approach for the hydraulic calculations with a modular approach for all other calculations. This combination yields models that can be very fast, even for very large models and yet have sufficient accuracy. The accuracy of the model can be increased by using smaller time steps in the model, but that comes at the price of speed. An optimized approach that is commonly made is to run the hydraulic calculations at a smaller time step than the modular calculations. The justification is that changes in composition and equilibrium constants happen much slower than changes in pressures and flows. In this way the most time-consuming calculations are executed less frequently and this benefits the overall speed of the model. Most commercial dynamic simulators use some form of this hybrid approach.

2.6 PROCESS SIMULATION BEST PRACTICES

This part will describe best practices for steady-state simulations, although some of the best practices apply equally well to dynamic models. The key message is to not blindly start modeling but to consider first what needs to be modeled and what is the purpose of the model.

2.6.1 Chemical Components and Thermodynamic Models

2.6.1.1 Component Lists

The first task is to select the components to model the system; the goal should be to select as few components as possible as more components means a slower model.

The information at hand for creating a simulation may contain a definition of the feed composition with a list of 30 components. Yet 15 of those may have a mole fraction in the feed that is negligible. One needs to consider if any of the components seemingly present in negligible amounts will accumulate somewhere in the process in a more concentrated form.

As an example, the feed to a natural gas liquefaction plant is often pretreated using a distillation process to remove the heavier compounds. The amount of isopentane in the original feed may be less than a tenth of a percent, but in the bottom product of that distillation column, there is over 5% of that compound. However, if the scope of the model does not include this distillation, it is quite likely that the isopentane can be removed from the component list. Another obvious reason to keep a trace component in the component list is that one of the objectives of the study is the behavior of that trace component in the plant.

Components to be modeled in gas processing plants are mostly pure chemical compounds. However, the feed to the plant may contain heavier hydrocarbons represented as hypothetical or pseudocomponents. In most cases, all these heavier compounds can be lumped into a single pseudo-component or hypothetical component.

Considering the feed components alone is often not sufficient. Consider the whole process to be modeled and add components that may not appear in the main process feed, but are added elsewhere in the system. Water and ethylene glycol are likely candidates for this in a gas processing plant. Blanketing gas or air may also be required as part of the main component suite.

Components exclusively used in utility systems should be placed in a different component list. A typical example could be to have a component list with only water for use in modeling cooling water and steam systems.

2.6.1.2 Thermodynamic Model Selection

In general, for any process with pressures above 10 bar (145 psig) the gas phase should be considered to be nonideal and an EOS should be used to model the gas phase. In gas processing this is almost always the case. If the fluids to be modeled contain significant amounts of polar compounds, then the liquid phase should be modeled using an activity model or another specific thermodynamic model (electrolytes, for example). In gas processing, the main process seldom has a lot of polar compounds and hence the main process fluid is usually modeled entirely using an EOS like Peng Robinson or SRK. Small amounts of water (a polar compound) are usually handled with sufficient accuracy in these EOSs, at least the way these are implemented in commercial simulation packages.

The densities of liquids are usually not well modeled with EOSs; most commercial simulators will use a different liquid density method by default to overcome this.

Especially for the sections of the plant that work at cryogenic conditions, it is best to use Lee–Kesler enthalpies rather than the EOS enthalpies. This is not a default option in commercial simulators and is the responsibility of the user.

In gas processing there are typically two units that require a specific thermodynamic model. The removal of water using TEG or other solvents requires the use of a different model to accurately model the equilibrium between the solvent and water and also of contaminants like benzene, toluene, and xylenes. The removal of CO_2 and sulfur compounds also requires a specific model, often combined with specific equipment models.

2.6.2 The Simulation Model

As discussed in previous paragraphs, the level of detail of the model in each piece of equipment should fit the purpose of the model. This selection has a large impact on the solution speed. Before applying any of the suggestions provided, give some thought to that purpose of the model! It could be used only for a couple of calculations, if so, it does not pay to spend an hour improving the model. Balance the optimizations with the expected use of the model.

If all simulations were such that each piece of equipment is only solved a single time, the considerations around number of components, model detail, etc. to ensure sufficient model speed would probably be irrelevant. But, there are two factors in modeling that result in each equipment block being calculated many times over: recycles and iterative procedures to reach a specification. These iterative procedures include adjusts, controllers, design-specs, or something else depending on the modeling software used.

2.6.2.1 Model Speed and Iterations

The presence of physical recycles in a plant requires an iterative solution. These recycles mean that in a typical modular solution of the model, there are streams that must start with an estimated state. This estimation must then be updated until the estimated state matches within a user-defined tolerance with the calculated state. The number of iterations required varies, but is typically in the order of five iterations. Multiple recycles can be nested and if a particular equipment is inside three of these loops, that one block may be calculated a 125 times before the final convergence of the model. If that block is very simple and is solved in one-hundredth of a second, the impact on the total solution time is limited. If that block is complex and requires 1 s to solve, the solution of that block alone will take 2 min. So, it is important to minimize the number of recycles and also to put the tear streams or recycle blocks in locations that will ensure the fastest solution.

Some commercial software provides an automated way to determine these tear streams. These automatically determined tear streams are not necessarily the best choices. The algorithm will normally find the minimum number of tear streams, but the selected location may not be optimum for stability. For nonsequential modular solvers, there usually is no algorithm that can determine the number of recycle blocks or the best location. The user must make an analysis that depends on model topology and also the way the model constraints have been specified. The advantage here is that some recycles can be avoided.

The second factor that induces iterations in a model is the presence of control or adjust blocks. These blocks vary a temperature, flow, pressure, or another specification in the model to obtain the desired value of some property that cannot be specified directly in the model. Anytime such a block seems necessary, check and double check if it is really required. The block may not be needed because a direct specification is in fact possible, albeit with some effort. The block may not be needed because obtaining the specification may not be crucial for the model. An often observed example is the use of an iterative block to obtain a volumetric flow in terms of actual volume flow. Most simulators do not accept this as a specification, but a calculator block or spreadsheet can easily be used to obtain a fluid density and to calculate the required mass flow from this density and the desired volume flow.

Once the iterative procedure has been optimized, it may make sense to look at the nature of the calculations of each object in the iterative calculations. If a block makes a lot of complex calculations, it pays to check if the results of those complex calculations are really needed in each iteration.

The calculation of a heat exchanger can serve as an example. Assume that in the model the exchanger outlet temperature and pressure are fixed, but that a rigorous exchanger model is used to determine the allowable amount of fouling. That allowable amount of fouling may be calculated hundreds of times, and yet it is only the very last calculation that is useful. In this case, it may make sense to replace the rigorous calculation of the exchanger with a simple calculation and to have a parallel block that calculates the fouling that only executes once after everything else has been solved. All simulators have that sort of capability; you just need to find how it is done.

2.6.2.2 Solution Order
Once a model becomes complex with several recycles and adjusts or controllers, the order of solution becomes important. In many scenarios the wrong order of solution may result in nonconvergence. An obvious case could be the steady-state solution of a compressor with an antisurge loop. Consider that the compressor is solved using a performance map, it operates at fixed speed, and the net feed is such that it the compressor would operate below the minimum allowable feed flow. The amount of recycled material needs to be adjusted until the volumetric feed to the compressor equals the flow that matches the control line of that compressor.

In an initial case, assume that the recycle block or tear stream is located upstream of the compressor block (Fig. 2.1). If the solution sequence is set up such that the adjusting iteration executes before the recycling, there is a problem. The result of the modification of the amount recycled will not propagate to the feed of the compressor as long as the recycle is not solved. Therefore, to the adjusting block it will look as if changes to the amount recycled have no impact on the compressor feed flow.

In a second case the recycle is upstream of the location where the amount recycled is specified (Fig. 2.2). In this second case, even though the order of solution will not prevent convergence, it may still influence the number of iterations required.

2.6.2.3 Model Robustness
No model will only be used to look at the results of just one set of problem specifications, but often a model seems to be made only for that initial set of

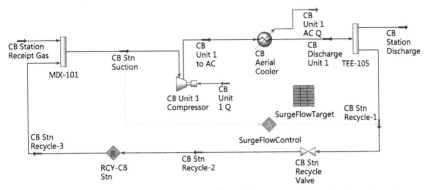

Figure 2.1 Compressor antisurge process flow diagram, adjusted value downstream of recycle. *CB*, compressor block; *RCY*, recycle.

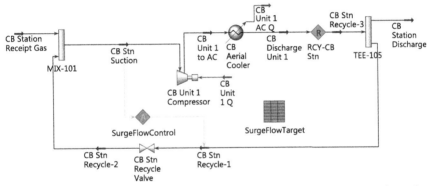

Figure 2.2 Compressor process flow diagram, adjusted value downstream of recycle. *CB*, compressor block; *RCY*, recycle.

specifications. When one or more of those specifications change value, the model may not solve anymore.

A simple example is a model of a distillation column where the specification is the desired flow rate of the overhead vapor. If the feed rate to the model is increased, the column will most likely still converge, but if the feed rate is reduced to 50% of the original rate, it is quite likely that the column will no longer converge. It may still be possible to achieve convergence, but some of the initial estimates would require adjustment and the operating model of the column may be radically different from the initial operating mode. This obviously begs the question if that new operating mode is the intention or not. It is far more likely that the intended operation of the column would see the specified overhead flow reduced to 50% as well when the feed is reduced. So, a specification of the overhead flow as a fraction of the feed flow would probably be a better and also more robust specification.

Another example is a model of a burner that takes some fuel source and has air or another oxygen source added to it. The mixer is then modeled in a reactor model to reproduce the effects of the combustion. A typical adjustment or control in the model is a variation of the flow of the oxygen source to achieve a concentration of oxygen in the flue gas. A properly configured control or adjust block would have limits to the flow that can be used. If the fuel flow is changed by a factor 10 or so, it is likely that the limits in the adjust block will no longer be valid. Even if they are, it is possible that the convergence will take a long time because the air flow must be changed by a factor of 10 as well. A more robust solution would be to create a linear relationship between the fuel flow and the oxygen source flow and then adjust the multiplier in that relationship to reach the target oxygen content in the flue gas. With this setup, changes in fuel flow rate will require very few or possibly no iteration at all. If the nature of the fuel or the oxygen source is changed, more iterations will be required, but only the first time after that change.

In general, it is a good idea to use your understanding of the process to add feed-forward control—type calculations to facilitate convergence of the model.

2.7 CASE STUDIES

2.7.1 Gas Dehydration with TEG

A gas dehydration plant with TEG typically consists of an absorber column where lean TEG is fed at the top of the column and the wet gas feeds into the bottom of the column. The TEG that has absorbed the water is then called the rich TEG. This rich TEG is let down in pressure, heated, and regenerated to lean TEG in a distillation column. The regeneration has several variants that can be used depending on the desired water content of the lean TEG.

2.7.1.1 Thermodynamic Model Selection

At the very least, the selected model needs to properly model the vapor—liquid equilibrium of the main components considered: methane, water, and TEG. The EOS models typically used to model hydrocarbon systems cannot model this system accurately as both water and TEG are polar molecules with a more complex interaction. The most commonly used simulators make available a model that is specifically tuned for this application. It is best to be aware of the developments for such systems. As an example, years ago the recommended thermodynamic model for TEG dehydration was the Peng Robinson EOS. The model was adequate because modifications under the hood forced the behavior of the system to properly represent this equilibrium. When the focus of the industry also included the distribution of BTEX (benzene, toluene, ethyl-benzene, xylenes) in the process, this EOS model was no longer adequate. A new specific TEG dehydration package was developed to properly model the distribution of the BTEX and also better model other aspects such as the solubility of TEG in high-pressure methane.

2.7.1.2 Modeling the TEG Dehydration System

The main points of attention are the modeling of the columns and the modeling of TEG recycling.

The absorber column typically does not pose many issues for convergence. The main task is to choose a number of stages that is sufficient to dehydrate the gas. This will mainly depend on the initial water content of the gas and the desired final water content and lean TEG purity. The regenerator column mostly has a very low number of stages (one or two). Despite its relative simplicity, it is more difficult to converge the column because of the large temperature difference between the top of the column and the bottom. The key is to choose the correct specifications for the column. Usually the top of the column has a temperature specification or a reflux ratio specification. The bottom of the column often also has a temperature specification. The reason for the latter choice is that TEG starts degrading at temperatures above 400°F and hence 400°F is a typical bottom temperature specification. If a higher-purity TEG is required, other solutions like stripping gas are used rather than a higher bottoms temperature.

When modeling TEG recycling, it is important to make the model robust against TEG losses. Although TEG losses in the overhead of both columns are minimal, there are many "opportunities" for losses while running a simulation model. A temporary use of an alternative specification on the regenerator column might result in significant losses of TEG in the overhead. If the recycling system is not set up to cope with this, the model may not converge. The typical solution to this is to include a TEG makeup stream with a flow that is calculated to be just enough to maintain a specified circulation rate of the TEG. The specified circulation rate may be a fixed number, or in a design case it may be better to use a number calculated based on the amount of water that needs to be removed from the gas stream.

A dynamic model of this process would not pose special challenges once a steady-state model is available. With a dynamic model the column solution typically poses fewer issues as specifications are replaced by the actions of controllers on the condenser and reboiler duty. The time span over which a dynamic model is run is typically at most a couple of hours, the loss of TEG over that time span is very small and hence the model does not really need to represent the TEG makeup.

2.7.2 Sour Gas Sweetening with Amines

A sour gas sweetening process looks very similar to dehydration with TEG. It also consists of an absorber column where lean amine is fed at the top of the column and the sour gas feeds into the bottom of the column. The amine that has absorbed the sour components is then called the rich amine. This rich amine is let down in pressure, heated and regenerated to lean amine in a distillation column. Sour gas sweetening is a more challenging simulation problem because the interaction between the amine and the sour components is a chemical reaction and the amine solution can be catalogued as an electrolytic fluid.

Given the complexity of this system, it is a good idea to consider if a rigorous model is necessary or if a simple component splitter block will be good enough.

2.7.2.1 Thermodynamic Model Selection

The model selected needs to account for the electrolytic nature of the amine solution. It also needs to account for the chemical reactions that take place between the CO_2 and H_2S and the amine. Depending on the amine type used, the reaction rates may also come into play to model the system accurately. Select a thermodynamic package that is specific for gas sweetening and make sure it is capable of handling the particular amine or amine mixture modeled. An additional problem is that there exist myriad licensor processes that use amine mixtures or amines combined with activator components the exact nature of which is unknown.

2.7.2.2 Modeling the Absorption and Regeneration

The classic distillation unit operation in a simulator assumes vapor—liquid equilibrium on each stage and sometimes an efficiency is applied. In a gas-sweetening application, equilibrium is seldom reached and the approach to equilibrium is different for different components. The rigorous method to calculate the column is to use a rate-based approach. This approach calculates the rate of transfer of a component from one phase to the other based on convection, diffusion, and equilibrium. This requires the addition of the rate of reaction between the amine and the dissolved CO_2 and H_2S.

Several specialized packages exist for this purpose. Some of these are simulators specific for the purpose of gas sweetening and some of them are included in the more widely used process simulation packages. Given the complexity of the phenomena that are modeled, considerably more time may be spent to model these columns than for modeling a TEG dehydration column or a condensate stabilizer column, for example.

When trying to replicate the mass and energy balance of a licensor process, be aware that the reboiler duty of the regenerator column is extremely sensitive to the requested residual content of CO_2 or H_2S in the lean amine. It is typically better to impose the duty received from the licensor and to compare the sour component concentration than to impose the concentration. The resulting reboiler duty may be a multiple of the duty reported by the licensor if the residual CO_2 concentration is specified. When specifying the duty, the error on the CO_2 concentration may well be less than 5%.

When setting up the recycle of the amine, more care is needed than with TEG dehydration. Water and amines are lost at different rates; in a steady-state simulation it is best to calculate makeup rates for each such that the total desired circulation rate is maintained and the desired amine concentration is maintained. In a dynamic model, the control scheme of the makeup flows needs to replicate the real-life control scheme.

2.7.3 Turboexpander NGL Recovery

A turboexpander plant has a high degree of heat integration. The same type of heat integration is commonly found in all processing units that operate at subambient temperatures. The high-pressure feed gas is typically cooled in a multistream heat exchanger before being fed into an expander unit. The near isentropic expansion reduces the temperature enough to condense the heavier hydrocarbons in the gas. This condensate is removed in a knockout drum and the power generated by the expander is used to recompress the gas. This combination is called a turboexpander. A second compressor is needed to bring the gas back to the feed pressure and a column is used to produce NGL to the desired specification.

2.7.3.1 Thermodynamic Model Selection

An EOS like Peng Robinson or SRK is most often used to model this process. As the heat integration and compressor and expander power calculation is important here, the EOS enthalpy method is very often replaced with the Lee—Kesler method. A particular problem that may need addressing is the behavior of trace amounts of mercury in the process. The multistream heat exchangers are usually constructed in aluminum and this metal is very sensitive to the presence of mercury. EOSs do not properly represent the behavior of mercury out of the box, so check if the simulator has made special provisions for this.

2.7.3.2 Modeling the Multistream Exchanger

A dedicated multistream exchanger is usually available in the simulator. It is important to keep in mind a key design strategy used in multistream exchangers. The physical exchanger normally operates such that all the cold streams share the same temperature at a given location along the length of the exchanger and similarly for the hot streams. The multistream exchanger model typically includes this assumption in the way the calculations are done.

Consider the model in Fig. 2.3. Stream "Column OVHD" (Overhead) is warmer than stream (Knock–out) "Expander KO Out." The multistream model typically assumes that "Column OVHD" enters the exchanger in the middle at a location where the exchanger internal cold pass temperature matches the temperature of the stream. Similarly, it is assumed that if a cold stream has an outlet temperature different from the highest cold stream outlet, it exits the exchanger at the location that matches the specified outlet temperature.

A multistream exchanger can have multiple inlets and outlets to optimize the heat recovery. The resulting representation using a single multistream exchanger block can become difficult to understand. It is often useful to split the multistream exchanger into multiple blocks such that the

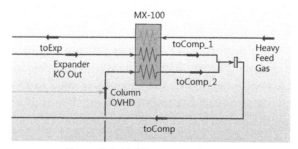

Figure 2.3 Multistream exchanger.

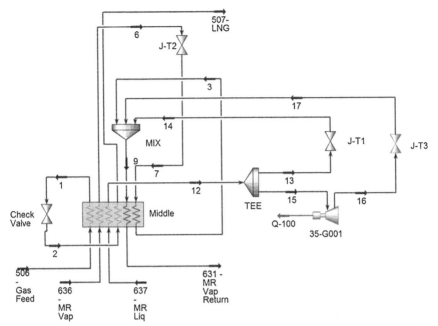

Figure 2.4 Single-block multistream exchanger. *LNG*, liquefied natural gas; *J-T*, Joule-Thompson.

process flow diagram (PFD) view matches the physical reality of where streams enter and leave the exchanger.

Splitting one exchanger block into three separate blocks introduces additional degrees of freedom. These degrees of freedom should first be consumed by specifications that impose the design strategy of having a common cold temperature and a common hot temperature everywhere in the exchanger. In the PFD for the split block exchanger (Refer to Figs. 2.4 and 2.5), the bottom and middle block should each have two specifications that set the temperature of an outgoing hot stream identical to one of the other outgoing hot streams and the top block should have one such specification.

2.7.3.3 Modeling the Turboexpander

For a pure design–type model, all simulators provide models for expanders and compressors and the power generated by the expander can in one way or another be transferred to the compressor. However, when it comes to modeling an existing plant, the modeling of the expander in rating mode is a lot more difficult using the standard available unit operations. The key problem is that an expander is more restricted in the feed flow rate it can accept. In a real plant, this is accomplished with a combination of actions on

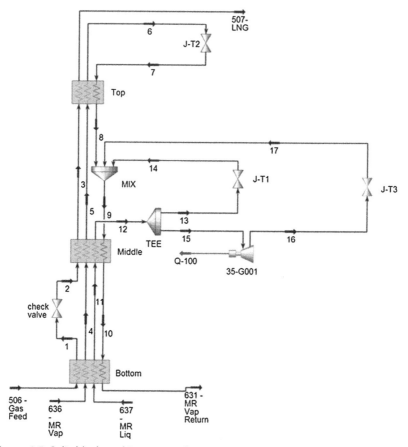

Figure 2.5 Split block multistream exchanger. *LNG,* liquefied natural gas; *J-T,* Joule-Thompson.

the inlet guide vanes and by using a bypass valve. In addition, the inlet pressure can be allowed to float to accommodate additional throughput. The performance characteristic of the expander is also often implemented as an inverted compressor curve where the true performance characteristic is quite different. The speed of the expander and the compressor are linked and a power balance needs to be done to determine the speed of the unit. Commercial add-on modules are available to model turbo expanders taking into account these limitations (Fig. 2.6).

2.7.4 MEG Regeneration

Ethylene glycol is one of the main products used to protect gas pipelines against hydrate formation. At the arrival facilities of a gas pipeline protected

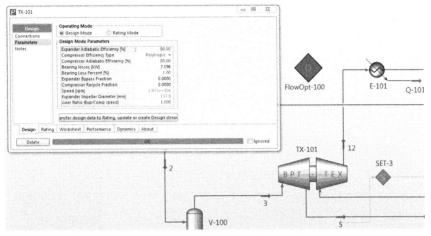

Figure 2.6 Commercial add-on to model a turboexpander.

with MEG there needs to be a unit to regenerate the rich MEG to lean MEG. The regeneration consists of a partial or total evaporation process. The partial evaporation only aims at removing the absorbed water from the MEG. However, in many cases the MEG picks up more than only water. Salts may be present in the water coming from various sources: salty formation water, corrosion inhibitor by-products, and corrosion products. These salts accumulate in the MEG and if the amount of salts is too high, the MEG must either be replaced or undergo a more thorough regeneration. The regeneration evaporates both water and MEG that are then separated in a distillation column.

2.7.4.1 Thermodynamic Model Selection

The challenges posed here are somewhat similar to the ones for TEG dehydration. However, an accurate model is not easy because special thermodynamics are necessary to handle the equilibrium between MEG, water, and various hydrocarbons. The Cubic Plus Association (CPA) model is capable of handling this mixture, but assure that the CPA model in the simulator also has the necessary parameters to do this accurately. The salts add another level of difficulty to this; there are a few specialist thermodynamic packages or modeling tools that can handle this.

2.7.4.2 Modeling the Process

The process itself does not pose particular problems. One issue sometimes encountered is that patents may cover the process and vendors may be secretive about the actual operating conditions. Some of the separation equipment like the separation of salts from concentrated MEG may not have a rigorous model; a simple component splitter-type model may be the only available option.

CHAPTER 3

Process Control

In this chapter, readers are introduced to a better understanding of the fundamental concepts in dynamics and process control, which include dynamic process characteristics, control system components, control algorithms, input—output relationships, open-loop and closed-loop characteristics, control loop tuning, and control strategies.

3.1 DYNAMIC PROCESS CHARACTERISTICS

The dynamic process characteristics of most processes can be described by three elements: resistance, capacitance, and dead time contributions. These elements will determine how the process responds to changes.

3.1.1 Resistance-Type Processes

The characteristic of a resistance element is the ability to transfer material or energy. Flow through a pipe is the most common example of a resistance-type process. A strictly resistance-type process has a proportional only response. Any change in the resistance, for example, a control valve opening, will result in an immediate proportional change in the flow. The amount of change is a function of the process gain. Any change in the load, for example, the upstream and downstream pressure in this case of flow through a pipe, will also result in an immediate proportional change in the flow. The amount of change is again a function of the process gain.

3.1.2 Capacitance-Type Processes

The characteristic of a capacitance element is the ability to store energy and mass. Thermal capacitance is simply the product of the mass of an object and its specific heat. Energy can be stored in a heat exchanger. Mass capacitance is the accumulation of mass. In other words, input does not always equal output. Separators and surge dampers are common examples of strictly mass capacitance processes.

The gas capacitance of a tank is constant. The liquid or solid capacitance equals the cross-sectional area of the tank at the surface. If the cross-sectional area is constant, then the capacitance is constant at any head.

Modeling, Control, and Optimization of Natural Gas Processing Plants
ISBN 978-0-12-802961-9
http://dx.doi.org/10.1016/B978-0-12-802961-9.00003-6

An example of a strictly capacitive process is a tank containing liquid with liquid removed at a constant rate, for instance, a positive displacement pump. The change in level of the tank is determined by the product of the difference between inlet and outlet flow rate and the time that the tank has been filling divided by the capacitance, or cross-sectional area, of the tank. Since the capacitance of many tanks can change with level and therefore time, for instance, a sphere or horizontal tank with elliptical heads, a differential describes the capacitance.

Large vessels in relation to the flows have a larger capacitance and hence the more slowly the controlled variable changes for a given change in the manipulated variable. The capacitance of the process tends to attenuate disturbances and hence makes control less difficult. The capacitance of the process is measured by its time constant. The time constant may be calculated from the differential equations used to model the process, but as an approximation is roughly equal to the process residence time.

3.1.3 Process Dead Time

Another contributing factor to the dynamics of many processes is a transportation lag or dead time. The definition of dead time is the delay in time for an output to respond to an input. For example, if the flow into a pipeline is increased, a period of time elapses before the outlet flow changes. This elapsed time is a function of the line length and fluid velocity. Dead time is caused by the time to move material from one point to another, so it is frequently referred to as "transportation lag." Some examples of dead time include static mixers and conveyor belts. Dead time also occurs when energy or mass sources must flow to the destination where temperature, composition, flow, or pressure is sensed. Delay also arises when a control signal requires time to travel between two points.

Dead time interferes with good control inasmuch as it represents an interval during which the controller has no information about the effect of a load change or control action already taken. When designing a control system, every attempt should be made to minimize this delay by properly locating instruments and sampling points, insuring sufficient mixing, and minimizing transmission lags.

Dead time is particularly difficult for feedback controllers for a variety of reasons. Feedback controllers are designed for immediate reaction and dead time varies with throughput.

3.1.4 Inertia-Type Processes

The motion of matter as described by Newton's second law typifies inertia-type processes. These effects are important when fluids are accelerated or decelerated as well as in mechanical systems with moving components.

3.1.5 Combinations of Dynamic Characteristics

Few processes are strictly resistance or capacitance types. Systems that contain a resistance element and a capacitance element result in a single time constant process. As noninteracting resistance and capacitance elements are added to the system, multiple time constants result. When introducing inertial effects or interactions between first-order resistance and capacitance elements the processes exhibit a second-order response.

Single time constant processes can be described with a first-order differential equation containing a constant gain and a time constant. The gain is independent of the time characteristics and is related to the process amplification or ultimate output divided by input. The time constant describes the time required for the system to adjust to an input. This time constant is simply the product of the resistance and the capacitance.

A first-order process response without dead time is described with the following equation:

$$\text{Change in Output} = \text{Gain} \times \text{Change in Input} \times \left(1 - e^{-t/\text{time constant}}\right)$$

$$(3.1)$$

Two distinct principles apply to all first-order processes without dead time. First, the initial rate of change is the maximum and if unchanged the system would reach the ultimate output in a period of time equal to the time constant. Second, the actual response is exponential with the output reaching 63.2% of the ultimate value at a time lapse equal to the time constant of the system. Dead time just shifts this response to begin after the delay has elapsed.

Combinations of first-order lag with dead time adequately estimate the response of most processes for purposes of simulation and control (Fig. 3.1).

3.1.6 Examples: Simple Systems

3.1.6.1 Vessels and Piping

Systems of vessels with connecting piping are simple examples of combined capacitance and resistance processes. The piping contributes a resistance element while the vessels contribute a capacitance element to the system. First-order dynamics accurately describe these combinations.

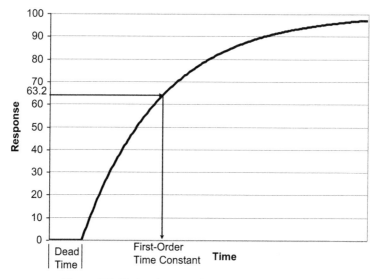

Figure 3.1 First-order open-loop process response.

Consider a large gas tank that dampens flow rate and pressure variations as depicted below:

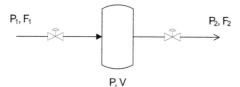

The volume of the tank determines the process gain between vessel pressure and gas flow rates at constant temperature and composition. The time constant can also be determined from the following:

- Volume of the vessel (V)
- Molecular weight of the gas (MW)
- Inlet gas rate (F_1)
- Outlet gas flow rate (F_2)
- Ideal gas constant (R)
- Temperature (T)
- Pressures (P)

In this case, the time constant is (Liptak, 1995):

$$2 \cdot V \cdot MW / (R \cdot T \cdot (F_1 + F_2)/2) \cdot (P_1 - P_2) \qquad (3.2)$$

For:
- A vessel with volume of 100 cubic feet
- Gas with molecular weight of 29
- A destination pressure of 100 psia (P_2)
- A source pressure of 105 psia (P_1)
- Gas constant (10.732 psia-ft^3/lbmole-°R)
- Gas temperature of 62.33°F (522°R)
- Gas inlet and outlet flow rate of 10 lb mass per minute (F_1 and F_2)

The time constant for the change of vessel pressure will be about 0.5 min or 30 s.

For source gas temperature and composition variations, the length of pipe introduces a dead time.

3.1.6.2 Heat Exchangers

Heat exchangers are common in gas processing plants. There are six PVs involved:

1. Cold stream flow
2. Hot stream flow
3. Cold stream inlet temperature
4. Cold stream outlet temperature
5. Hot stream inlet temperature
6. Hot stream outlet temperature

Four of the variables are independent, whereas two are dependent. The overall system considerations usually dictate the flow rate of one stream and the inlet temperatures. This leaves the outlet temperatures and one flow rate as potential independent variables. Most often, an outlet temperature is the controlled variable, whereas an inlet flow rate or bypass flow is the manipulated variable.

Assuming that the temperature change of the tube wall dominates the transfer function, then the mass of the tubes and the specific heat of the tubes divided by the product of the flow rates and specific heats of the two process streams approximates the time constant. The following empirical formula for the time constant can be derived (Liptak, 1995):

$$K \cdot M_t \cdot Cp_t / (F_w \cdot Cp_w \cdot F_c \cdot Cp_c) \qquad (3.3)$$

For the case of:
- M_t (mass of tubes) is 4400 lbm
- Cp_t (specific heat of tubes) is 0.112BTU/lb°F
- F_w (flow of warm stream) is 55 lb/s

- Cp_w (specific heat of warm stream is 0.70 BTU/lb°F
- F_c (flow of cold stream) is 80 lb/s
- Cp_c (specific heat of cold stream is 0.80 BTU/lb°F

where K = 100 for English units.

Then, the time constant will be 20 s.

The dynamic considerations for furnaces and boilers are similar to heat exchangers.

3.1.6.3 Pipelines

Pressure or flow at the end of the line is the typical control objective for a pipeline. Pressure at the beginning of the pipeline is independent as determined by a gas source or a compressor. The source pressure and the controlled pressure determine the flow rate.

The static gain (G) of flow to upstream pressure (P_1) is:

$$G = F \cdot P_1 / \left(P_1^2 - P_2^2\right) \tag{3.4}$$

where P_2 is the downstream pressure.

The sum of the dead time and time constant can be determined empirically as (Liptak, 1995):

$$K \cdot V \cdot MW \cdot (P_1 - P_2) / T \cdot Z \cdot F \tag{3.5}$$

where K is 0.015 for English units; V is the volume of the pipeline in cubic feet; MW is the molecular weight of the gas; P is the pressure in pounds per square inch absolute; T is the gas temperature in Kelvin; Z is the compressibility factor; F is the flow in pounds per second.

For the following conditions of:

- F = 1500 lb/s
- P_1 = 2000 psia
- P_2 = 1000 psia
- V = 3,226,200 cubic feet (approximately 48 inches diameter by 50 miles pipeline)
- MW = 20
- T = 50°F (10°C or 283K)
- Z = 0.95

The process gain will be 1 lb/s/psi and the dead time plus first-order time constant will be 2400 s or 40 min.

Pumps, compressors, and turbines have static relationships between the PVs, so constant speed rotating equipment introduces no dynamics to the process. Variable speed equipment usually introduces negligible lag.

3.1.6.4 *Effects of Variable Conditions*

In the examples given earlier the time constant depends on flow rate, which varies. Therefore the process gain, time constant, and dead time will vary with flow. Flow control valves can also introduce nonlinear responses due to changes in inlet or outlet pressure as well as the characteristics of the valve itself.

Instrumentation can also introduce gain variations. Temperature measurements typically introduce a time lag, whereas composition analyzers normally introduce a dead time. Pneumatic control valves without positioners can introduce delays. Although digital control systems significantly reduce the delays experienced with analog systems, delays can occur due to scan frequency.

3.2 CONTROL SYSTEM COMPONENTS

Proper choice of controls requires knowledge of the requirements of the process and the corresponding characteristics of the controls available. The control mode determines the controller characteristics. The controller mode ultimately chosen depends on the process dynamics and control objectives.

Feedback or closed-loop controllers compare a measurement (controlled variable) with its desired value (set point) and generate a correction signal (change in the controller output) to a control element (manipulated variable) based on the difference between the measurement and the desired value. Negative-feedback controllers act to eliminate the difference. A controller can operate on the difference between the measurement and the desired value, the error, in many different ways (Fig. 3.2).

Control action is determined by the way the controller outputs dependence on the error. In direct action, the controller output signal increases

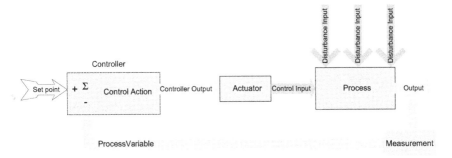

Figure 3.2 Process control loop.

when the controller input exceeds the set point. Conversely, with reverse action, the controller output signal decreases when the controller input rises.

Controller, valve, and process must be matched by the correct choice of action. The control loop and the valve failure position must be determined first because these two decisions dictate the correct controller action.

There are two basic modes of control:

- Discrete (On—Off)—The valve is either fully open or fully closed, with no intermediate state. Alternatively, a manipulated variable such as a motor may be turned "on" or "off."
- Continuous—The valve can move between fully open or fully closed, or be held at any intermediate position.

Some common modes of continuous control responses include proportional, integral (reset), derivative, and combinations of these modes.

3.3 CLOSED-LOOP CONTROL

The most important aspect of control is the dynamic relationship between the measured, dependent, or controlled variable and the correcting, independent, or manipulated variable. Control of the desired outcome of the measured variable requires knowledge of the response of the corrective action as a function of time.

The dynamic behavior of a closed-loop system depends on the gain or transfer function of the process and the gain or transfer function of the control system including the influences of valve movements and control signal delays. The size and configuration of the equipment such as piping, vessels, heat exchangers, distillation columns, compressors, and pumps, as well as the laws of chemistry and physics, establish the process gain.

Knowledge of the process gain is necessary to provide stable control of the loop. Systems of differential equations can accurately calculate the process gain. The solution of the differential equation system with properly defined initial conditions provides the values of the PVs over time.

In the absence of a dynamic simulation, the first-order lag with dead time model will approximate the process gain.

3.3.1 On—Off Control

The most simple control strategy and consequently the least expensive in terms of initial cost and maintenance is the on—off control. A room thermostat is an example of on—off control. In the field, on—off control is often called snap acting. Level controls on low-pressure separators and temperature controls on indirect fired heaters are often snap acting. The controller will turn the manipulated variable on when the measurement deviates from set point and will turn the manipulated variable off when the measurement is at set point. In practical applications a small error, or dead band, is allowed to eliminate constant cycling of the manipulated variable (Shinskey, 1996).

In many cases, the irregular value of the controlled or manipulated variable resulting from a snap-acting control creates serious operating and control problems downstream of the processes. In these cases, proportional control is used (Seborg, 2011).

3.3.2 Proportional Control

Proportional (P) control is the basic action employed in all controllers not using snap action. Proportional control means that the change in the controller output (m) is proportional to the change in the error (e). The constant magnitude of the change in the controller output to the change in error is the controller proportional gain (K_c).

$$m = K_c \cdot e + b \tag{3.6}$$

A bias (b) adjusts the output when required; for instance, 4 mA for electronic loops that measure 4—20 mA. There is no bias when the controller output is expressed as 0—100%.

The proportionality constant is sometimes expressed in terms of percent proportional band (P_b) where:

$$P_b = 100/K_c \tag{3.7}$$

In some control systems, proportional action is adequate to meet the control objectives. Proportional control will result in deviations from the set point except when the set point is at 50% of span. The amount that the measurement is off the set point is the offset. In fact, offset is the difference between the measured value of the controlled variable and the set point. By increasing the gain the offset reduces, but it is not eliminated. However, most processes become unstable unless the time constant is extremely high (very slow processes).

3.3.3 Integral Control

Integral (I) action is added to the controller to minimize or eliminate offset. Integral control means by the property that the change in the controller output is proportional to the integral of the error. The prime purpose of integral is therefore to prevent offset and keep the controlled variable at the control point even as the process load changes. The integral algorithm constantly calculates the accumulated error and corrects for past over- or undercompensation with an integral time setting (T_i).

$$m = 1/T_i \cdot \int e \, dt + b \qquad (3.8)$$

The integral time expressed in repeats per minute or minutes per repeat is sometimes called the reset time, representing the time (in minutes) for the integral action to repeat.

Integral only control is sometimes used in flow control systems and is suitable for pure time delay processes. There are conditions when an integral controller will continue to accumulate error even when it is not desirable and the output will go to the maximum or minimum value. Some examples are inactive processes such as a surge controller does not detect surge conditions or a cascade master is switched off cascade. An external reset or antireset wind-up feature is necessary to protect against saturation in the idle state.

Integral action introduces a lag into the system and is usually employed with proportional mode. These modes are commonly combined into a single algorithm where the proportional gain is applied to the integral error as well as the current error. The resulting algorithm is:

$$m = K_c \cdot \left(e + 1 \Big/ T_i \cdot \int e \, dt \right) + b \qquad (3.9)$$

The integral rate requires careful adjustment. When integral rate is set correctly, valve movement occurs at a rate to which the process can respond. If set too fast, cycling will result because the valve moves faster than the process and the measurement, and cycling results. If set too slow, the process will not recover quickly enough resulting in sluggish control.

Integral action lags proportional action. The proportional action provides the quick response to correct for the upset, whereas the integral provides the gradual correction to bring the controlled variable back to the set point.

Proportional—integral control has the advantage of reducing the initial error that occurs with integral only while eliminating offset.

3.3.4 Derivative Control

Proportional plus integral control does not provide correction that is rapid enough for certain processes. Derivative (D) response may be added to anticipate a change in process load and transmit a corrective signal to minimize the lag. This action corrects based on the rate of change of the deviation from the set point.

Derivative action leads proportional in that it causes the valve to move faster and further than it ordinarily would with proportional action alone. Derivative action is normally specified only for temperature control systems, which are characterized by large process and measurement lags or for special applications such as antisurge systems where a rapid valve response is imperative.

Derivative control means that the change in the controller output (m) is proportional to the rate of change in the error (e). The derivative mode predicts errors and takes corrective action before occurrence of the error proportional to the derivative time (T_d).

$$m = T_d \cdot de/dt \qquad (3.10)$$

The derivative time is the duration of time that the algorithm will look into the future. Larger derivative times will contribute larger corrective action. This corrective action will also protect against overshoot of the set point.

As the size of processing equipment increases, the momentum makes it difficult to control without derivative action. However, there are several limitations of derivative control. When the change in the error is constant, derivative mode takes no action. In addition, derivative will react to set point changes and step changes in discontinuous measurements, such as chromatographs. For these reasons, derivative control is never used alone. In a few instances, proportional-derivative control may be used, such as special instances of slave controllers in a temperature cascade system and batch pH control.

3.3.5 Proportional-Integral-Derivative Control

Although proportional with integral (PI) control eliminates offset, considerable time may occur before the error returns to zero. Proportional, integral, and derivative (PID) controllers are used in processes with large capacitance, or long time constants, or slowly changing process outputs. Temperature and concentration loops are the most common examples.

PID modes are typically combined into one algorithm.

$$m = K_c \cdot \left(e + 1/T_i \cdot \int e \, dt + T_d \, de/dt \right) + b \qquad (3.11)$$

This algorithm can simultaneously respond to current error, eliminate offset, and anticipate error. Much research has been conducted to determine optimum settings for proportional gain, integral time, and derivative time.

Sample and hold algorithms are useful when the dead time of a loop is greater than the time constant. Conventional PID algorithms are not adequate in these instances. This algorithm utilized the standard PID algorithm part of the time by switching the controller between automatic and manual. The period for which the controller switches to manual is set by a timer to exceed the dead time and then alternates to automatic for an output update. To affect the required magnitude of change, the integral setting must be increased.

The most common form of PID controllers includes:

1. Flow
2. Pressure
3. Liquid level
4. Temperature
5. Composition

3.3.6 Advanced Control

This section will discuss some advanced control methods including:
- Feedforward control
- Cascade control
- Interaction and decoupling
- Nonlinear control
- Adaptive control
- Model–based control
- Model predictive control
- Optimizing control
- Self-tuning controllers

3.3.6.1 Feedforward Control

Feedforward control differs from feedback control in that the load or primary disturbance is measured and the manipulated variable is adjusted so that deviations in the controlled variable from the set point are minimized. The controller can then reject disturbances before they affect the controlled variable. For accurate feedforward control, steady-state or dynamic analysis should be the basis for models that relate the effect of the manipulated and disturbance variable on the controlled variable. Since the model is an approximation and not all disturbances are measured, feedforward control should always be used in conjunction with feedback control. This combination will allow compensation for measured and unmeasured disturbances as well as model mismatch (Seborg, 2011).

3.3.6.2 Cascade Control

Processes that respond to disturbances with long time delays or lags are difficult to control with a single feedback controller (Seborg, 2011). One relatively simple way to improve the dynamic response is to use a secondary measurement and controller. The secondary measurement and control should recognize the effect of a disturbance before the primary controlled variable is affected. Cascade control where the primary controller, also called master controller, output becomes the set point for the secondary controller, also called slave controller, is a readily configured strategy in most computer control systems.

Cascade control is commonly used on distillation towers for composition control. The primary measurement of a key component typically has a time delay and feedback control is difficult. Temperature of a tray that is sensitive to compositional changes can be used as a secondary measurement. The tray temperature usually responds faster than the analyzer.

In the case of distillation composition control, the response can be further improved by configuring the heat medium flow to a reboiler or reflux rate as a secondary control for the appropriate tray temperature control. As a rule of thumb, the primary loop should be five times faster than the secondary loop.

3.3.6.3 Override and Selectors

One approach to addressing more controlled variables than manipulated variables is to use a selector (Seborg, 2011). The most common types of selective control are high and low selectors. This allows control action based on a selection criterion for multiple measurements.

Override is another type of selective control. In this case, a PV that reaches a predetermined high or low limit dictates the controller objective.

3.3.6.4 Interaction and Decoupling

An interactive system exists when a manipulated variable for one controlled variable affects the controlled variable of another loop. Additional feedback loops occur between unpaired manipulated and controlled variables in these cases and destabilize the system (Shinskey, 1996).

Relative gain analysis reveals interactions and the degree of interaction. Pressure and level are functions of flow differences, whereas temperature and composition are function of flow ratios. These differences and ratios can often be used to decouple and minimize interactions. This technique is rational decoupling. Decoupling is similar to feedforward control except the load variable replaces a manipulated variable.

When ratios and differences are not effective in eliminating interactions, then the controllers can be detuned to restore stability. This detuning will lead to loss in control performance.

Another technique includes linear decouplers calculated from process gains. The issues with this approach are constraints and initialization. When a manipulated variable is constrained, then two controllers compete for an unconstrained variable. One controller will wind down, whereas the other controller winds up. Decoupling using measured values of the manipulated variables overcomes the initialization and constraint issues. Even with these enhancements, decouplers are difficult to match and remain matched with the process leading again to stability issues.

Partial decoupling can be used to stabilize interactive loops. The best practice is to protect the least likely controlled variable to change set point, the slowest or the most important controlled variable.

Other techniques for decoupling are an adaptive multivariable controller on feedforward loops and dynamic matrix methods. Subsequent sections describe these methods.

Blending and distillation systems often present interactive systems that require decoupling for effective control.

3.3.6.5 Nonlinear Control

Conventional process control systems utilize linear dynamic models. For highly nonlinear systems, control techniques directly based on nonlinear models provide significantly improved performance.

Most real processes display some nonlinear behavior (Seborg, 2011). The process gain and dead time can change with load, time with equipment degradation, and dead time with transportation lag. In many cases, linear controllers provide adequate control performance. As the degree of nonlinearity increases, improved control performance may be necessary and desired.

There are two classes of nonlinear control: discontinuous and continuous. The discontinuous methods include on−off and three state devices. These discontinuous methods are adequate only when accurate regulation is not essential. Continuous nonlinear control methods include fuzzy logic, output filtering, characterization, and dead-band or gap action.

3.3.6.6 Adaptive Control

Programmed and self-adaptive controls comprise the two classes of adaptive controls. Programmed adaptive applies when a measured variable yields a predictable response from the control loop. Programmed adjustments are continuous or discontinuous including gap action and switching controller gains. Adaptive methods include self-tuning, model reference, and pattern recognition (Shinskey, 1996).

An example of variable, predictable gain on a control loop occurs in plug flow situations commonly found in heat exchangers. At low flow the dead time, steady-state gain, and dominant time constant are proportionately higher than at high flow. For instance, at 50% flow, these parameters are all twice as high as at 100% flow.

The equation for a flow-adapted PID controller is:

$$m = K_c \cdot \left(w \cdot e + w^2/T_i \cdot \int e \, dt + T_d \cdot de/dt \right) + b \qquad (3.12)$$

where K_c, T_i, and T_d are the PID settings at full-scale flow and w is the fraction of full-scale flow.

Gap action, mentioned in the nonlinear control section, can also be adaptive. A dead band is placed around the error used in the calculation of the PID action. No action is taken within the dead band. Level control of surge tanks, pH control, and other systems where too much control action may introduce instability are typical examples.

Gain scheduling is another form of nonlinear control or programmed adaptation. Variable breakpoint control is another description of this type of control. This method is often similar to gap action. Table 3.1 shows a comparison:

Another common method of controller gain adjustments is an error squared formulation. This produces a nonlinear controller with similar characteristics to the typical implementation of gain scheduling. Little action is taken at small deviations from set point compared with the action

Table 3.1 Comparison of gap action to gain scheduling

Situation	Gap action	Gain scheduling
Close to set point	No action within a predetermined dead band	Typically very little action with low controller gain
Further from set point	One set of predetermined PID tuning parameters used outside the dead band	One or more additional sets of predetermined PID tuning parameters used with higher gain the further the departure from set point

taken at larger deviations. This method should be used with caution and limits placed as instability may occur at large deviations from set point.

When unknown or immeasurable disturbances impact the control loop response, programmed adaptation is not possible. Industry practices several forms of self-adaptive methods including self-tuning regulator, model reference adaptive, and pattern recognition.

All self-tuning systems have common elements of an identifier, controller synthesizer, and controller implementation. Types of self-tuning regulators include dead beat, stability theory, fuzzy logic, pole–placement, generalized predictive, and minimum variance algorithms (Shinskey, 1996).

The system identifier estimates the parameters of the process. It determines the response of the controlled variable due to a change in the manipulated variable. These changes may be deliberate or may be transients that normally occur.

The synthesizer calculates the controller parameters based on a control objective. Recursive estimators determine the optimal PID gains. A function block updates the controller parameters calculated at a predetermined frequency or subject to heuristics.

Model reference adaptive methods are classified as optimal or response specification-type algorithms. Minimum variance types employ a least squares method comparing the PV to its set point. The minimum variance adaptive controllers often become unstable when the time delay varies or is unknown. Generalized predictive control uses predictive horizon techniques to overcome the time delay issues experienced with minimum variance methods.

Pole placement and pole assignment routines are common response specification-type adaptive controllers. These algorithms use a desired closed-loop frequency to determine preferred controller parameters. This technique requires an excitation of the process normally in open loop.

Pattern recognition does not use a model to self-adapt a controller. When a change in set point occurs, the dead time, sensitivity, and steady state gain are determined from a positive and negative load response. After identification of the measurement noise the oscillation, damping ratio, and overshoot are evaluated for desired parameters based on integrated absolute error (IAE). Neural network algorithms are also effective for pattern recognition. The radial basis function form of neural networks is ideal due to their ability to analyze time series data.

Optimization can be attempted with single loop controllers using several techniques. However, multivariable algorithms are much more effective for optimization solutions.

Evolutionary optimization and gradient search methods are two approaches with single loop controllers (Shinskey, 1996). With evolutionary optimization, online experimentation is conducted to determine the optimum. An independent variable is changed and the response of the optimization objective measured to determine whether there is an improvement. If the response has a negative impact on the objective, then the independent variable will be moved in the opposite direction. If the response has a positive impact on the objective, then the independent variable will continue movement in the same direction. If there is no response, then it is assumed that the optimum has been found.

An example is minimization of a cold separator for a cryogenic natural gas liquids recovery operation. An inlet gas split ratio can be increased or decreased and the response of the cold separator temperature monitored. If an increase in the split ratio causes an increase in the cold separator temperature, then the split ratio will be decreased. This decrease will continue until there is no change in the cold separator temperature or until the temperature experiences an increase.

Gradient search methods including the steepest ascent algorithms are sometimes implemented for optimizing control. Complications arise when a constraint is encountered. Typically, imposition of penalty terms on the objective function for active constraints force control back within a feasible region.

3.3.6.7 Internal Model Control

The Internal Model Control (IMC) philosophy relies on the International Model Principle, which states that control is only achievable if an implicit or explicit representation of the process describes the control system (Shinskey, 1996). In particular, if the control scheme is based on an exact model of the process, then perfect control is theoretically possible. IMC uses open-loop step response Laplace transfer functions with process gain and time constant to predict a measurement change due to a change in the set point. The IMC philosophy can also be used to generate settings for conventional PI or PID controllers. The algorithm allows for a model to process difference and biasing the set point to remove the steady-state offset, which becomes a feedback adjustment method. The biased set point is filtered to obtain a reference trajectory. A single tuning factor filters the time constant. High tuning factors lead to gentle control action, whereas low tuning factors lead to aggressive control action.

IMC models are inherently stationary and linear. Like PID controllers, IMC must be tuned for changes in process gain or time constant. The model should be reparameterized when the process dynamics change substantially (Shinskey, 1996).

3.3.6.8 Model Predictive Control

Model Predictive Control (MPC) has become the most popular advanced control technique for difficult control problems. The main idea of MPC is to choose the control action by repeatedly solving on line an optimal control problem. This aims at minimizing a performance criterion over a future horizon, possibly subject to constraints on the manipulated inputs and outputs, where the future behavior is computed according to a model of the plant. Model predictive controllers rely on dynamic models of the process, most often linear empirical models obtained by system identification. The models predict the behavior of dependent variables (i.e., outputs) of a dynamical system with respect to changes in the process independent variables (i.e., inputs) (Seborg, 2011).

If a reasonably accurate dynamic model of the process is available or can be derived, then the model updated with current measurements can be used to predict the future process behavior. The values of the manipulated inputs are calculated so that they minimize the difference between the predicted response of the controlled outputs and the desired response.

Despite the fact that most real processes are approximately linear within only a limited operating window, linear MPC approaches are adequate in the majority of applications with the feedback mechanism of the MPC compensating for prediction errors due to structural mismatch between the model and the plant. In model predictive controllers that consist only of linear models, the superposition principle of linear algebra enables the effect of changes in multiple independent variables to be added together to predict the response of the dependent variables. This simplifies the control problem to a series of direct matrix algebra calculations that are fast and robust.

MPC offers several important advantages (Seborg, 2011):

1. The manipulated inputs are adjusted based on how they affect the outputs,
2. Inequality constraints on PVs are easily incorporated into the control calculations,
3. The control calculations can be coordinated with the calculation of optimum set points, and
4. Accurate model predictions provide an indication of equipment problems.

MPC is widely considered the control method of choice for challenging multivariable control problems and normally provides very quick payout.

Identification of the dynamic process model is key to the successful implementation of MPC techniques. In most MPC applications, special plant tests are required to develop empirical dynamic models from input—output data. Although these plant tests can be disruptive, the benefits from implementing MPC generally justify this short-term disruption to normal plant production.

3.4 DEGREES OF FREEDOM

In developing the control system for a process it is desirable to know precisely the number of PVs, which the system designer is entitled to attempt to regulate, commonly known as the degrees of freedom of the process. An analysis of the dynamics of a system must begin with an analysis of the degrees of freedom. The degrees of freedom are simply the number of variables that describe the system less the independent relationships that exist among the variables. The degrees of freedom dictate the number of variables that must be specified to properly simulate a system. Once specified, the design of a control structure requires knowledge of the control degrees of freedom. This is the number of controllable variables. It is very easy to calculate this number, even for quite complex processes, because it is equal to the number of manipulated variables (the number of control valves in the process). For control system design, the number of automatic controllers is limited by the degrees of freedom. However, these variables are different than the design optimization variables. When optimizing, the degrees of freedom determine the number of variables that must be included in the objective function to yield a single solution (Seborg et al., 2011).

Once a variable is determined to be a manipulated or specified variable then it becomes an independent variable and uses a degree of freedom. Other variables that have a relationship to this variable are dependent variables.

3.5 CONTROL LOOP TUNING

Controller tuning, which is the adjustment of the controller parameters to match the dynamic characteristics (or personality) of the entire control loop, has been referred to as the most important, least understood, and most poorly practiced aspect of process control. The tuning of feedback controllers is part of the overall commissioning of plants and machines. Controller tuning is misunderstood and often "trial and error" must be used to achieve an acceptable combination of the tuning parameters for a particular process. "Good" control is a matter of definition and depends upon such factors as individual preference, process disturbances and interactions, product specifications, etc.

This section discusses control quality and the performance criteria to consider when tuning a controller. It also examines methods of tuning PID controllers and the considerations for selection of the PID settings.

3.5.1 Quality of Control

If the PID controller parameters (the PID terms) are chosen incorrectly, the controlled process input can be unstable, i.e., its output diverges, with or without oscillation, and is limited only by saturation or mechanical breakage. Tuning a control loop is the adjustment of its control parameters (gain/proportional band, integral gain/reset, derivative gain/rate) to the best values for the desired control response.

Good process control begins in the field, not in the control room. Sensors and measurements must be in appropriate locations and valves must be sized correctly with appropriate trim. The final control elements, such as control valves, execute the changes required to manipulate the preferred process parameters like flow, temperature, pressure, level, ratio, etc. If the instruments in the field do not function as required, then one cannot expect the overall process control to perform optimally. Tuning should be revisited as the process and equipment change or degrade. The controllability of a process depends upon the gain that can be used. Higher gain yields the greater rejection of disturbance and the greater the response to set point changes. The predominate lag is based on the largest lag in the system. The subordinate lag is based on the dead time and all other lags. The maximum gain that can be used depends on the ratio of the predominate lag to the subordinate lag. From this we can draw two conclusions: (1) decreasing the dead time increases the maximum gain and the controllability and (2) increasing the ratio of the longest to the second longest lag also increases the controllability. In general, for the tightest loop control, the dynamic controller gain should be as high as possible without causing the loop to be unstable.

Controller tuning involves setting the three constants in the PID controller algorithm to provide control action designed for specific process requirements. The response of the controller is described in terms of the responsiveness of the controller to an error, the degree to which the controller overshoots the set point and the degree of system oscillation. Note that the use of the PID algorithm for control does not guarantee optimal control of the system.

The best response to a process change or set point change varies depending on the application. Some processes must not allow an overshoot of the PV beyond the set point if, for example, this would be unsafe. Other processes must minimize the energy expended in reaching a new set point. Generally, stability of response (the reverse of instability) is required and the process must not oscillate for any combination of process conditions and set points. Some processes have a degree of nonlinearity and so parameters that work well at full-load conditions do not work when the process is starting up from no load. This section addresses some of the performance criteria that are used in the tuning of a controller, including overshoot, decay ratios, and error performance.

3.5.2 Controller Response

The three possible general extremes of response are: overdamped, critically damped, and underdamped. There are several common performance criteria that can be used for controller tuning, which are based on the characteristics of the system's closed-loop response. Some of the more common criteria include overshoot, offset, rise time, and decay ratio. These criteria may be classified as peak related and time related.

3.5.2.1 Peak-Related Criteria

Fig. 3.3 shows a set point step response plot with labels indicating peak features:

- A = size of the set point step
- B = height of the first peak
- C = height of the second peak

    The popular peak-related criteria include:

- Peak overshoot ratio (POR) = B/A
- Decay ratio = C/B

    In the plot, the process variable (PV) was initially at 20% and a set point step moves it to 30%. Applying the peak-related criteria by reading off the PV axis:

- A = (30 − 20) = 10%
- B = (34 − 30) = 4.0%
- C = (31 − 30) = 1%

    So for this response:

- POR = 4.0/10 = 0.40 or 40%
- Decay ratio = 1/ 4.0= 0.25 or 25%

    Many control engineers believe that for most control loops the optimum tuning is a 1/4 wave decay. This response provides robust disturbance

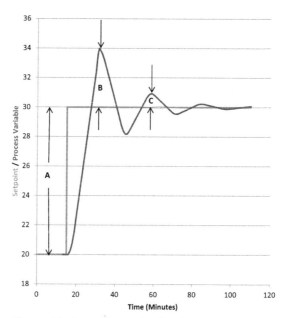

Figure 3.3 Controller peak-related response criteria.

rejection. The quarter decay ratio has proved through experience to provide a good trade-off between minimum deviation from the set point after an upset and the fastest return to the set point.

However, some control loops require a fast response but no overshoot control performance.

No overshoot means no peaks, and thus, $B = C = 0$. This means the peak-related criteria are not useful as performance comparison measures.

3.5.2.2 Time-Related Criteria

An additional set of measures focus on time-related criteria. Fig. 3.4 is the same set point response plot but with the time of certain events labeled.

The clock for time-related events begins when the set point is stepped, and as shown in the plot, include:

- Rise time = time until the PV first crosses the set point
- Peak time = time to the first peak
- Settling time = time to when the PV first enters and then remains within a band whose width is computed as a percentage of the total change in PV (or ΔPV).

The 5% band used to determine settling time in the plot was chosen arbitrarily. Other percentages are equally valid depending on the situation.

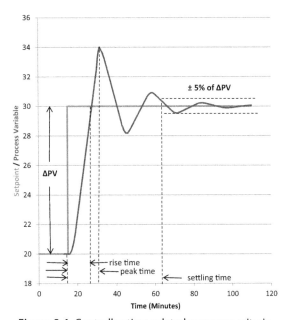

Figure 3.4 Controller time-related response criteria.

From the plot, we see that the set point is stepped at time t = 15 min. The time-related criteria are then computed by reading off the time axis as:
- Rise time = (26 − 15) = 11 min
- Peak time = (30 − 15) = 15min
- Settling time = (61 − 15) = 46min for a ±5% of ΔPV band

3.5.2.3 When There Is No Overshoot
We should recognize that the peak and time criteria are not independent; a process with a large decay ratio will likely have a long settling time and a process with a long rise time will likely have a long peak time.

In situations where no overshoot is desired there is no POR, decay ratio, or peak time to compute. Even rise time, with its asymptotic approach to the new steady state, is a measure of questionable value.

In such cases, settling time, or the time to enter and remain within a band of width we choose, still remains a useful measure.

Fig. 3.5 shows the identical process as that in the previous plots. The only difference is that in this case, the controller is tuned for a moderate response.

We compute for this plot:

Settling time = (61 − 15) = 46 min for a ±5% of ΔPV band.

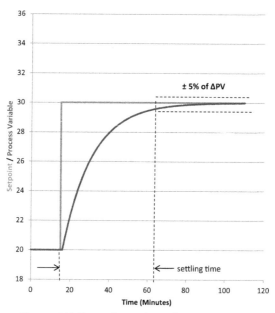

Figure 3.5 Controller with moderate response.

3.5.3 Error Performance Criteria

Simple performance criteria, such as decay ratio, use only a few points in the response and are simple to use. However, more complicated error performance criteria are based on the entire response of the process. Several criteria or objectives have been proposed. Among the most popular are minimum integral of square error, minimum integral of absolute error (IAE), and minimum integral of time and absolute error.

The integrated square error criterion uses the square of the error, thereby penalizing larger errors more than smaller errors. This gives a more conservative response, i.e., faster return to the set point. In mathematical terms, with e representing the error as a function of time, we can write:

$$ISE = \int_0^\infty e(t)^2 \, dt \qquad (3.13)$$

IAE essentially takes the absolute value of the error.

$$IAE = \int_0^\infty |e(t)| \, dt \qquad (3.14)$$

The integrated time absolute error criterion is the integral of the absolute value of the error multiplied by time. ITAE results in error penalties existing over time, even though they may be small, which results in a more heavily damped response. The mathematical expression for this criterion follows:

$$ITAE = \int_0^\infty |e(t)| \cdot t \, dt \qquad (3.15)$$

3.5.4 Tuning Methods

There are a number of methods for tuning single-loop controllers. A few methods will be described, which are based on simple experiments or simple models, and do not require any frequency domain analysis (although such analysis may enhance understanding of the resulting closed-loop behavior).

Many techniques are useful for tuning control loops including: experience and a sense for the adequacy of control, heuristics, complex mathematics, and self-tuning systems. Control loop tuning may be accomplished in closed loop where the controller is set on automatic or open loop where the controller is set on manual.

3.5.4.1 Process Reaction Curve Methods

Sometimes a step response curve is called a "reaction curve." In the process reaction curve methods a reaction curve is generated in response to a change. This process curve is then used to calculate the gain, integral time, and derivative time of the process. These methods are performed in open loop, so no control action occurs and the process response can be isolated.

To generate a process reaction curve, allow the process to reach steady state or as close to steady state as possible. Then, in open loop to eliminate any control action, introduce a small disturbance and record the reaction of the PV.

Methods of process analysis with forcing functions other than a step input are possible, and include pulses, ramps, and sinusoids. However, step function analysis is the most common, as it is the easiest to implement.

3.5.4.1.1 Ziegler—Nichols Open-Loop Procedure

Ziegler and Nichols (1942) developed controller tuning equations based on field measurements of the ultimate gain and ultimate period. It consists of the two following steps:

1. Determination of the dynamic characteristics, or personality, of the control loop
2. Estimation of the controller tuning parameters that produce a desired response for the dynamic characteristic determined in the first step, in other words, matching the personality of the controller to that of the other elements in the loop.

In this method the dynamic characteristic of the process are represented by the ultimate gain of a proportional controller and the ultimate period of oscillation of the loop.

For a manual tuning test, the derivative time is set to zero and the integral time is set at least 10 times larger than normal so that most of the controller response is from the proportional mode. The controller gain is then increased to create equal sustained oscillations. The controller gain at this point is the ultimate gain and the oscillation period is the ultimate period. In industry, the gain is only increased until decaying oscillations first appear to reduce the disruption to the process. Autotuners and adaptive controllers can eliminate the need for manual controller tuning. The relay or on—off method is used by autotuners to automatically compute the ultimate period and gain by switching the controller output when it crosses and departs from a noise band centered on the set point (Oglesby, 1996; Dahlin, 1968).

Table 3.2 Tuning parameters for the open-loop
Ziegler–Nichols method

	Gain	Reset	Derivative
P	$\frac{\tau}{K\alpha}$	—	—
PI	$\frac{0.9\tau}{K\alpha}$	$0.3/\alpha$	—
PID	$\frac{1.2\tau}{K\alpha}$	$0.5/\alpha$	0.5α

The ultimate gain for self-regulating processes per the amplitude ratio is (Dahlin, 1968):

$$K_u = \left[1 + (\tau_1 \cdot 2 \cdot \pi/T_u)^2\right]^{0.5}/K_o \qquad (3.16)$$

For the Ziegler–Nichols ultimate oscillation method, the controller gain is a fraction of the ultimate gain and the integral time is a fraction of the ultimate period as follows for a PI controller.

The Ziegler–Nichols open loop recommended controller settings for the quarter decay ratio are given in Table 3.2. ΔM is the change in controller output, Y is the change in the PV, α (min) is time until the intercept of the tangent line and original process value, and τ (min) is the time until the intercept of the tangent line and final process value less α as shown in Fig. 3.6. K is the ratio of Y to ΔM.

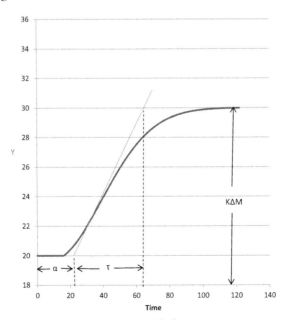

Figure 3.6 Ziegler–Nichols parameters.

The Ziegler—Nichols tuning method was one of the first formal methods that found wide adoption. This method is also known as the "reaction curve" method. To use the Ziegler—Nichols method, the process must be stable. With the controller in manual, change the output by a small amount outside the noise band. Monitor the process to observe the effect and verify the validity of the dynamic process data collected.

3.5.4.1.2 Cohen—Coon Tuning Method

Cohen and Coon (1953) modified the Ziegler—Nichols open loop tuning rules. The modifications are insignificant when the dead time is small relative to the time constant, but can be important for large dead time. The Cohen—Coon tuning parameters are shown in Table 3.3.

As with the Ziegler—Nichols open-loop method recommendations, the Cohen—Coon values should be implemented and tested in closed loop and adjusted accordingly to achieve the quarter decay ratio.

3.5.4.1.3 Internal Model Control Tuning Rules

IMC tuning rules have proved to be robust and yield acceptable performance when used in the control of common processes. In general, analytical IMC tuning rules are derived for PI/PID compensators by matching an approximate process model to a low-dimensional reference model. The IMC controller structure depends on two factors: the complexity of the model and the performance requirements stated by the designer.

Since the general IMC method is unnecessarily complicated for processes that are well approximated by first-order dead time or integrator dead time models, simplified IMC rules were developed by Fruehauf et al. (1994) for PID controller tuning as given in Table 3.4.

The IMC-PID tuning rules are sometimes applied in industry. However, the widely published IMC tuning rules, while providing adequate suppression of output disturbances, do a poor job suppressing load

Table 3.3 Tuning parameters for the Cohen—Coon method

	Gain	Reset	Derivative
P	$\frac{\tau}{K\alpha}\left[1 + \frac{\alpha}{3\tau}\right]$	—	—
PI	$\frac{\tau}{K\alpha}\left[0.9 + \frac{\alpha}{12\tau}\right]$	$\alpha\left[\frac{30 + 3\alpha/\tau}{9 + 20\alpha/\tau}\right]$	—
PID	$\frac{\tau}{K\alpha}\left[1.33 + \frac{\alpha}{4\tau}\right]$	$\alpha\left[\frac{32 + 6\alpha/\tau}{13 + 8\alpha/\tau}\right]$	$\alpha\left[\frac{4}{11 + 2\alpha/\tau}\right]$

Table 3.4 Simplified IMC rules

	$\tau/\alpha > 3$	$\tau/\alpha > 3$	$\alpha < 0.5$
Gain	$\tau/(2K\alpha)$	$\tau/(2K\alpha)$	τ/K
Reset	5α	τ	4
Derivative	$\leq 0.5\alpha$	$\leq 0.5\alpha$	$\leq 0.5\alpha$

disturbances when the process dynamics are significantly slower than the desired closed-loop dynamics. Morari et al. (1996) proposed to address this problem by including an additional integrator in the output disturbance while performing the IMC design procedure. This method provided adequate load disturbance suppression for many processes and has been applied to MPC (Morari, 1996). However, the resulting controllers do not have a PID structure.

3.5.4.2 Constant Cycling Methods
3.5.4.2.1 Ziegler–Nichols Closed-Loop Method
This tuning method is probably the most well-known tuning method. It is based on a simple closed-loop experiment, using proportional control only. The proportional gain is increased until a sustained oscillation of the output occurs (which neither grows nor decays significantly with time). The proportional gain that gives the sustained oscillation, and the oscillation period (time) are recorded. The proposed tuning parameters can then be found in Table 3.5.

In most cases, increasing the proportional gain will provide a sufficient disturbance to initiate the oscillation. Note that for controllers giving positive output signals, i.e., controllers giving output signals scaled in the range 0−1 or 0%−100%, a constant bias must be included in the controller in addition to the proportional term, thus allowing a negative proportional term to have an effect. Otherwise, the negative part of the oscillation in the plant input will be disrupted, which would also effect the oscillation of the output.

Table 3.5 Tuning parameters for the closed-loop Ziegler–Nichols method

	Gain	Reset	Derivative
P	$0.5\ K_u$	—	—
PI	$0.45\ K_u$	$1.2/T_u$	—
PID	$0.6\ K_u$	$2/T_u$	$T_u/8$

The steps required for the closed-loop Ziegler—Nichols methods are:
1. Place the controller into automatic with low gain and no reset or derivative.
2. Gradually increase the gain by making small changes in the set point until oscillations start.
3. Adjust the gain to force the oscillations to continue with a constant amplitude.
4. Note the gain (ultimate gain, K_u) and period (ultimate period, T_u.)
5. The ultimate gain, K_u, is the gain at which the oscillations continue with a constant amplitude.

Essentially, the tuning method works by identifying the frequency for which there is a phase lag of 180 degrees. For the tuning method to work, the system to be controlled must have a phase lag of 180 degrees in a reasonable frequency range and a gain that is large enough that the proportional controller is able to achieve a loop gain of 1 (0 dB). These assumptions are fulfilled for many systems.

The tuning method can lead to ambiguous results for systems with a phase lag of 180 degrees at more than one frequency. Despite its popularity, the Ziegler—Nichols closed-loop tuning rule is often considered to give somewhat aggressive controllers.

3.5.4.3 Autotune Variation Technique

The objective of autotuning methods is to obtain a PID controller capable of satisfying typical requirements such as rapid following, zero steady-state error, and overshoot suppression by means of a practical and robust method. The autotune method has several advantages over open-loop pulse testing methods, where (1) no prior knowledge of the system time constant is necessary, and (2) the method is closed loop tested, so the process will not drift away from the set point.

Relay feedback and frequency domain by magnitude and phase calculation autotuning are some of the most common methods to adjust the parameters of a PID controller. Relay feedback autotuning is used to obtain the ultimate frequency. This frequency is a reference and the design operating frequency will be a fraction of the reference.

Åstrom and Hägglund (1984) proposed an attractive method for determining the ultimate frequency and ultimate gain commonly called autotune variation (ATV). This method consists of an approximative method called harmonic balance method based on relay feedback autotuning. Fig. 3.7 shows a process transfer function with a feedback ideal

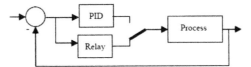

Figure 3.7 Block diagram of a relay autotuner. *PID*, proportional, integral, and derivative.

relay. In parallel with a feedback relay there is a PID controller, the achieved virtual controller.

An approximate condition for oscillation can be determined by assuming that there is a limit cycle with period T_u and frequency $w_u = 2\pi./T_u$ is determined such that the relay output is a periodic symmetric square wave. If the relay amplitude is d, a simple Fourier series expansion of the relay output shows that the first harmonic component has the amplitude $4d/\pi$. Assume that the process dynamics correspond to a low-pass system and that the first harmonic contribution dominates the output. The error signal then has the amplitude.

An adaptive controller has been achieved with the ATV autotuning technique or relay feedback autotuning to supply the frequency of the limit cycle and its associated gain, known as ultimate frequency and ultimate gain (Åström, 1989).

3.5.5 PID Tuning Software

Most modern industrial facilities no longer tune loops using the manual calculation methods discussed earlier. Instead PID tuning and loop optimization software are used to ensure consistent results. These software packages will gather the data, develop process models, and suggest optimal tuning. Some software packages can even develop tuning by gathering data from reference changes.

Mathematical PID loop tuning induces an impulse to the system and then uses the controlled system's frequency response to determine the PID parameters. In loops with response times of several minutes, mathematical loop tuning is recommended, because trial and error can take much time to find a stable set of loop values. Finding optimal values is even more time consuming to find with manual methods. Some digital loop controllers offer a self-tuning feature, which sends very small set point changes to the process, thereby allowing the controller itself to calculate optimal tuning values.

When generating dynamic process data, it is important that the change in the controller output signal causes a response in the measured PV that clearly dominates the measurement noise. One way to quantify the amount of noise in the measured PV is with a noise band. A noise band is based on the standard deviation of the random error in the measurement signal when the controller output is constant and the process is at steady state. Plus or minus three standard deviations of the measurement noise around the steady state of the measured PV (99.7% of the measurements are contained within the noise band) is conservative when used for controller tuning.

When generating dynamic process data, the change in controller output should cause the measured PV to move at least 10 times the size of the noise band. In other words, the signal to noise ratio should be greater than 10. For instance, if the noise band is 1°F, then the controller output should be moved enough during a test to cause the measured exit temperature to move at least 10°F.

With today's data historian capabilities, much high frequency information can be captured for calculating tuning factors. Trending capabilities are also quite helpful for these analyses.

Many PID tuning software packages include analysis of the entire control loop to detect valve or sensor issues. Stiction and hysteresis are two types of valve problems that a good software package can detect. These algorithms will also identify valves that are incorrectly sized.

Many of these packages will measure the entire valve travel over a period. This measurement highlights valves that will most likely require frequent maintenance due to poor tuning.

3.5.6 Choosing a Tuning Method

The choice of method will depend largely on whether or not the loop can be run in open loop for tuning and the response time of the system (Table 3.6).

If the system operates in open loop, then the best tuning method often involves subjecting the system to a step change in input, measuring the output as a function of time and using this response to determine the control parameters.

If the system must remain online, one tuning method is to initially set the I and D values to zero. Increase the P until the output of the loop oscillates, then the P should be left set to be approximately half of that value for a quarter amplitude decay-type response. Next increase I until any offset is correct in sufficient time for the process. Too much I will cause instability. Finally, increase D, if required, until the loop is acceptably quick to

Table 3.6 Comparison of tuning methods

Method	Advantages	Disadvantages
Ziegler–Nichols	Proven method. Closed-loop method	Process upset, some trial and error, very aggressive tuning
Tune by feel	No math required. Closed-loop method	Erratic, not repeatable
Software tools	Consistent tuning. Closed-loop or open-loop method. May include valve and sensor analysis. Allow simulation before downloading	Some cost and training involved
Cohen–Coon	Good process models	Some math required. Open-loop method. Only good for first-order processes

reach its reference after a load disturbance. Too much D will cause excessive response and overshoot. A fast PID loop tuning usually overshoots slightly to reach the set point more quickly; however, some systems cannot accept overshoot, in which case an "over damped" tune is necessary. This will require a P setting significantly less than half that of the P setting causing oscillation.

3.5.6.1 Flow Loops

Flow loops are too fast to use the standard methods of analysis and tuning.
   Analog versus digital control:
- Some flow loops using analog controllers are tuned with high gain.
- This will not work with digital control.

   With an analog controller, the flow loop has a predominate lag (L) of a few seconds and no subordinate lag.

   With a digital controller, the scan rate of the controller can be considered dead time. Although this dead time is small, it is large enough when compared with L to force a low gain. Table 3.7 gives typical PID settings for different types of control loops.

3.5.7 Lambda Tuning

Another class of tuning method that has become popular with the increased use of computing power is lambda tuning. Lambda tuning refers to all tuning methods where the control loop speed of response is a selectable tuning parameter. The closed-loop time constant is referred to as "lambda." Lambda tuning originated with Dahlin (1968); it is based on the same IMC

Table 3.7 Typical PID settings for different type loops

Loop type	PB (%)	Integral (min/rep)	Integral (rep/min)	Derivative (min)	Valve type
Flow	50–500	0.005–0.05	20–200	None	Linear or modified percentage
Liquid pressure	50–500	0.005–0.05	20–200	None	Linear or modified percentage
Gas pressure	1–50	0.1–50	0.02–10	0.02–0.1	Linear
Liquid level	1–50	1–100	0.1–1	0.01–0.05	Linear or modified percentage
Temperature	2–100	0.2–50	0.02–5	0.1–20	Equal percentage
Chromatograph	100–2000	10–120	0.008–0.1	0.1–20	Linear

theory as MPC (Morari, 1989; Chien, 1990) is model based and uses a model inverse and pole-zero cancellation to achieve the desired closed-loop performance.

The lambda tuning equations were developed for simplicity and practicality. For self-regulating processes the equations are as follows:

$$\lambda = \lambda_f \cdot \tau_1 \tag{3.17}$$

$$T_i = \tau_1 \tag{3.18}$$

$$K_c = T_i / (K_o \cdot (\lambda + \tau_d)) \tag{3.19}$$

$$T_d = \tau_2 \tag{3.20}$$

where K_c = controller gain, K_o = open-loop gain, λ = lambda (closed-loop time constant), λ_f = lambda factor, τ_d = total loop dead time, τ_1 = largest open-loop time constant, τ_2 = second largest open-loop time constant, T_i = integral time setting, and T_d = derivative time setting.

The lambda factor is the ratio of closed-loop time constant to the open-loop time where the open-loop time constant is the largest time constant (τ_1). For maximum load rejection capability, a lambda equal to the total loop dead time ($\lambda = \tau_d$) can be used and the loop will still be stable.

Lambda tuning is a model-based method with the goal of matching the set point response to a first-order time constant called lambda. Given a model, the tuning method for an ideal type PID controller is simple once you convert the units properly. Parallel- and series-type controllers require different tuning.

The design concept behind lambda is to cancel the process with the controller. The response is first delayed by the process dead time and then a first-order filter is used to obtain the desired response.

Once the field devices have been checked and corrected as required, an open-loop response test with the controller in manual is performed to understand the dynamics of the process. Testing should be performed over a range of normal operating conditions. The collected data should be fit to a simple dynamic model such as first order plus dead time and integrator plus dead time.

A unification of lambda, IMC, and Ziegler—Nichols reaction curve and ultimate oscillation tuning methods has been achieved. The controller tuning equations from diverse methods reduce to a common form, where the maximum controller gain is proportional to the time constant to dead time ratio (τ_1/τ_d) and is inversely proportional to the open-loop gain (K_o), commonly known as the process gain. This common form is easy to remember and provides insight as to the relative effects of process dynamics on tuning and hence on loop performance (Boudreau, 2006).

If a small lambda is substituted into Eq. (3.19), the equation reduces to the Simplified Internal Model Control (SIMC) equation, which has been documented to provide the best load rejection (tightest control) for self-regulating processes (Skogestad, 2003).

$$K_c = 0.5 \cdot \tau_1 / (K_o \cdot \tau_d) \qquad (3.21)$$

Eq. (3.21) was used by Ziegler and Nichols. The multiplier ranges from 0.4 to 0.8 and the exponent of the time constant to dead time ratio varies from 0.9 to 1.0. The differences in the multipliers or exponents are not important, because the maximum gain and the effect of errors or changes commonly seen in the identified process gain, time constant, and dead time is larger than the effect of the coefficients.

The definition of pseudo or "near" integrator gain is:

$$K_i = K_o / \tau_1 \qquad (3.22)$$

Substitution of Eq. (3.22) into Eq. (3.21) yields the SIMC tuning shown to provide the tightest control of integrating processes (Skogestad, 2003).

$$K_c = 0.5 / (K_i \cdot \tau_d) \qquad (3.23)$$

If the integrating process gain is defined as the reaction rate or the slope of the tangent line in Fig. 3.6 and the dead time is the delay time $(\tau_d = \alpha)$, then this is the controller gain per the Ziegler—Nichols method.

The "reaction curve" method is suitable for self-regulating processes with large time constants and integrating processes. However, this method is valid only when there is no change in the controlled variable before the controller output is changed so that the slope of the tangent line in Fig. 3.6 can be determined from just the ramp rate of the process variable after the change. Since the controller is in manual, this may not be the case. The change in ramp rates before and after the change should be used as a percentage.

The pseudo or "near" integrator gain becomes:

$$K_i = (CV_2/\Delta t - CV_1/\Delta t)/\Delta CO \tag{3.24}$$

where CV_1 = controlled variable before the change in controller output (%); CV_2 = controlled variable after the change in controller output (%); ΔCO = change in controller output to compensate for disturbance (%); Δt = module execution time (seconds).

Correction of the observed dead time for the effect of final element resolution or dead band may be required for tuning a controller for slow processes (McMillan, 2005).

Identification of the second largest time constant for the lambda method may be difficult. For a concentration, temperature, or gas pressure response, about half of the total loop dead time originates from the second largest time constant (T_2) and can be approximated by:

$$\tau_2 = 0.5 \cdot \tau_d \tag{3.25}$$

This IMC computation of the derivative time may be useful. Since in this case the time constant is much larger than the dead time, the dead time term in the denominator of Eq. (3.26) becomes just twice the largest time constant $(2 \cdot \tau_1)$. After simplification, the derivative time becomes approximately equal to half the dead time. Eq. (3.25) coupled with Eq. (3.26) reduces to Eq. (3.20).

$$T_d = \tau_1 \cdot \tau_d/(2 \cdot \tau_1 + \tau_d) \tag{3.26}$$

The lambda tuning equations for integrating processes are as follows:

$$\lambda = \lambda_f/K_i \tag{3.27}$$

$$T_i = 2 \cdot \lambda + \tau_d \tag{3.28}$$

$$K_c = T_i/\left(K_i \cdot (\lambda + \tau_d)^2\right) \tag{3.29}$$

The open-loop time constant is the inverse of the integrating process gain $(1/K_i)$. Since the lambda factor is the ratio of closed-loop time constant to open-loop time, lambda is equal to the total loop dead time ($\lambda = \tau d$). The loop will still be stable if the dynamics are accurately known.

In this case, Eq. (3.29) reduces to Eq. (3.23) but with a multiplier of 0.75 instead of 0.5. The dynamics are seldom known accurately enough to take advantage of this increase in the controller gain.

For many processes with a true or near an integrating response, the controller gains computed by Eq. (3.23) are too high when the disturbances are slow. A much lower controller gain can lead to undesirable oscillations with a long period. To prevent this from occurring, Eq. (3.30), which is developed from the transfer function for the closed-loop response of an integrating process, can be used to ensure the response is overdamped.

$$T_i > 4/K_i \cdot K_c \tag{3.30}$$

For self-regulating processes, the product of the controller gain and the process gain is much larger than integrating loops because the time constant is large, which means you can cancel out the product of gains in the numerator. If Eq. (3.22) is then used to obtain an equivalent process integrating gain, Eq. (3.31) reduces to Eq. (3.30).

The integral time to ensure an overdamped response in self-regulating processes is:

$$T_i > 4 \cdot (K_o \cdot K_c \cdot \tau_1)/(1 + K_o \cdot K_c)^2 \tag{3.31}$$

The "relay method" is extensively employed by "on-demand" auto-tuners to automatically compute the ultimate period and gain by switching the controller output when it crosses and departs from a noise band centered on the set point (McMillan, 2005).

The ultimate gain for self-regulating processes per the amplitude ratio is (Blevins, 2003):

$$K_u = \left[1 + (\tau_1 \cdot 2 \cdot \pi/T_u)^2\right]^{0.5}/K_o \tag{3.32}$$

where K_u = ultimate gain; T_u = ultimate period.

For the Ziegler–Nichols ultimate oscillation method, the controller gain is simply a fraction of the ultimate gain and the integral (reset) time is a fraction of the ultimate period as follows for a PI controller:

$$K_c = 0.4 \cdot K_u \tag{3.33}$$

$$T_i = 0.8 \cdot T_u \tag{3.34}$$

Considering the squared expression in the numerator of Eq. (3.32) is much larger than one yields Eq. (3.35). For most temperature and composition loops, the ultimate period is approximately four times the dead time ($T_u = 4 \cdot \tau_d$). Substitution of Eq. (3.34) and use of Eq. (3.32) yields Eq. (3.36). After multiplication of numerical factors, Eq. (3.36) becomes Eq. (3.37) (Eq. (3.21) with a slightly larger multiplier).

$$K_c = 0.4 \cdot ((\tau_1 \cdot 2 \cdot \pi)/T_u)/K_o \qquad (3.35)$$

$$K_c = 0.4 \cdot (\tau_1 \cdot 2 \cdot \pi)/(4 \cdot \tau_d)/K_o \qquad (3.36)$$

$$K_c = 0.6 \cdot \tau_1/K_o \cdot \tau_d \qquad (3.37)$$

If the ultimate period is about four times the dead time ($T_u = 4 \cdot \tau_d$), then the integral time ends up as about three times the dead time for the ultimate oscillation method, which is the same result obtained per Eq. (3.28) for the lambda tuning method when lambda is reduced to equal the dead time. This reset time is generally considered to be too fast. The SIMC method states that four times the dead time provides the best performance and an increase to eight times the dead time provides better robustness. If four times the dead time is used, then Eqs. (3.20) and (3.25) result in a reset time that is eight times the rate time setting. Although most of the literature shows the rate time is equal to 1/4 the reset time, in practice a rate time that is 1/8−1/10 of the reset time provides a smoother response.

The IAE can be derived from the response of a PI controller to a load upset (McMillan, 1991). The module execution time (Δt) is added to the reset or integral time (T_i) to show the effect of how the integral mode is implemented in some digital controllers. An integral time of zero ends up as a minimum integral time equal to the execution time so there is not a zero in the denominator of Eq. (3.38). For analog controllers, the execution time is effectively zero (Shinskey, 1993).

$$CO_{t2} - CO_{t1} = K_c \cdot (E_{t2} - E_{t1}) + [K_c/(Ti + \Delta t)] \cdot \int (E_t \cdot \Delta t) \qquad (3.38)$$

where E_{t1} = error between setpoint (SP) and PV before the disturbance; E_{t2} = error between SP and PV after complete compensation of the disturbance; E_t = error between SP and PV during the disturbance.

The errors before the disturbance (E_{t1}) and after the controller has completely compensated for the disturbance (E_{t2}) are zero ($E_{t1} = E_{t2} = 0$). Therefore the long-term effect of the proportional mode, which is the first term in Eq. (3.38), is zero. Eq. (3.38) reduces to Eq. (3.39) (McMillan, 1991).

$$\Delta CO = [(K_c/(T_i + \Delta t))] \cdot \int (E_t \cdot \Delta t) \qquad (3.39)$$

For an overdamped response:

$$IAE = \int (E_t \cdot \Delta t) \qquad (3.40)$$

where IAE = integrated absolute error from the disturbance.

The open–loop error is the peak error for a step disturbance for the case where the controller is in manual (loop is open). The open–loop error (E_o) is the open–loop gain (K_o) times the shift in controller output (ΔCO) required to compensate for the disturbance when the controller is auto (loop is closed).

$$E_o = K_o \cdot \Delta CO \qquad (3.41)$$

Eq. (3.39) solved for the IAE defined in Eq. (3.40) and the open–loop error defined in Eq. (3.41) becomes Eq. (3.42). If you ignore the effect of module execution time (Δt) on the integral mode, Eq. (3.42) for an overdamped response because the IAE is the same as the integrated error (E_i). Even for a slightly oscillatory response, the approximation has proved to be close enough (Shinskey, 1994).

$$IAE = \big(1/(K_o \cdot K_c) \cdot (T_i + \Delta t) \cdot E_o\big) \qquad (3.42)$$

For vessel temperature, concentration, and pressure control, we can use Eq. (3.21) for the maximum controller gain and four times the dead time for the minimum reset time to express the minimum IAE in terms of the process dynamics. The resulting equation is:

$$IAE = 2 \cdot (T_d/T_1) \cdot (4 \cdot T_d + \Delta t) \cdot E_o \qquad (3.43)$$

This equation can be independently derived by multiplying the peak error for a step disturbance by the dead time (Blevins, 2003; Shinskey, 1993).

In practice, controllers are not tuned this aggressively. Often the reset time is set equal to the time constant and a lambda factor of 1.0 is used, which corresponds to a controller gain that is about 10 times smaller than the maximum controller gain for a primary loop, such as composition, pressure, and temperature.

If the disturbance is not a step change, the IAE will be smaller. The effect of a slow disturbance can be approximated by adding the disturbance time constant to the open-loop time constant (τ_1) in the denominator of Eq. (3.43).

An increase in the module execution time shows up as an increase in the loop dead time for unmeasured disturbances. If the disturbance arrives immediately after the process variable is read as an input to the module, the additional dead time is about equal to the module execution time. If the disturbance arrives immediately before the process variable is read, the additional dead time is nearly zero. On average, the additional dead time can be approximated as 50% of the module execution time. Simulations that create a disturbance that is coincident with the controller execution will not show much of an effect of execution time on performance. This scenario misleads users into thinking the execution time of MPC is not important for load rejection. For chromatographs where the result is only available for transmission after the processing and analysis cycle, the additional dead time is 150% of the analyzer cycle time (McMillan, 2005; Blevins, 2003; McMillan, 1991).

Eq. (3.43) shows the effect of the largest time constant, loop dead time, and module execution time on absolute integrated error if the controller is always retuned for maximum performance. A detuned controller may not do much better than a tightly tuned controller for a larger loop dead time or module execution time (McMillan, 1991). Thus the value of reducing these delay times depends on the controller gain used in practice. For example, the controller gain is simply the inverse of the open-loop gain for a lambda factor of one in a loop with a dead time much smaller than the time constant. In other words, Eq. (3.19) reduces to Eq. (3.44).

$$K_c = 1/K_o \qquad (3.44)$$

Substituting Eq. (3.44) into Eq. (3.21) and solving for the dead time, you end up with Eq. (3.45). This shows a lambda factor of one on a primary reactor loop, which implies a dead time that is about $1/2$ of the time constant. The IAE for this case will not appreciably increase until the dead time is about $1/2$ of the time constant.

Thus time and money spent on reducing the dead time or module execution time below this implied dead time has little value unless the controller is retuned (McMillan, 1991).

$$\tau_d = 0.5 \cdot \tau_1 \qquad (3.45)$$

3.5.8 First Principles Process Relationships

Controller tuning settings may be based on first principle relationships to predict process cause and effects that can lead to improved controller tuning and performance. First principles predictions may also influence the choice of control valve trim and the feedforward design (McMillan, 2015).

Equations have been developed from first principle relationships for the process gains, dead times, and time constants of volumes with various degrees of mixing. The results show that for well-mixed volumes with negligible delays, the effect of flow cancels out for the controller gain if lambda is set equal to the dead time or the reaction curve method is used.

The effect of flow also cancels out for the reset time if the process is treated as a near integrator and the lambda integrating tuning rule is used, because the flow rate cancels out in the computation of the ratio of process gain to time constant that is the near integrator gain. This ratio and near integrator gain are inversely proportional to the process holdup mass.

For temperature control, the effect of changes in liquid mass cancels out because a change in level increases the heat transfer surface area covered. The integrator gain is nearly always proportional to the overall heat transfer coefficient that is a function of mixing, fluid composition, and fouling.

For gas pressure control, the equations show that liquid level has a pronounced effect on the process integrating gain for vessel pressure control because it changes the vapor space volume without any competing effect. The integrator gain for composition and gas pressure is inversely proportional to liquid level.

The equations also show that if the transport delay for flow injection is large compared with the time constant, which does occur for reagent injection in dip tubes for pH control, then the controller gain will be proportional to flow. Note that pH control is a class of concentration control.

For the control of temperature and concentration in a pipe, the process dead time and process gain are both inversely proportional to flow and the process time constant is essentially zero, which makes the actuator, sensor, transmitter, or signal filter time lag the largest time constant in the loop. Thus the largest automation system lag determines the dead time to time constant ratio.

For a static mixer, there is some mixing, and the process time constant is inversely proportional to flow but is usually quite small compared with other lags in the loop. The controller gain is generally proportional to flow for both cases.

Examples:

The integrating process gain (K_{ip}) for the control of liquid level by the manipulation of a flow:

$$K_{ip} = 1/\rho_o \cdot A_o \qquad (3.46)$$

The integrating process gain (K_{ip}) for the control of pressure per gas law by the manipulation of a flow:

$$K_{ip} = (R \times T_g)/V_g \qquad (3.47)$$

For the manipulation of feed temperature to control outlet temperature, the process gain (K_p) is:

$$K_p = (C_p \times F_f)/(C_p \times F_f - \Delta Q_r/\Delta T_o + U \times A) \qquad (3.48)$$

For the manipulation of feed flow to control outlet temperature, the process gain (K_p) is:

$$K_p = (C_p \times T_f)/(C_p \times F_f - \Delta Q_r/\Delta T_o + U \times A) \qquad (3.49)$$

where A_o is the cross-sectional area of liquid level; A is the heat transfer surface area; C_p is the heat capacity of process; F_f is the total feed flow; K_{ip} is the integrating process gain; K_p is the process gain; Q_r is the heat from reaction; R is the universal constant for ideal gas law; T_g is the gas temperature; T_o is the vessel outlet temperature; U is the overall heat transfer coefficient; V_g is the gas volume; ρ_o is the liquid density.

3.6 INDIVIDUAL UNIT OPERATION CONTROL AND OPTIMIZATION STRATEGIES

Natural gas processing operations are typically complex and require automatic control. This complexity may lead to more undesirable events and abnormal conditions. Most of the current gas processing plants, especially the smaller ones, have very few operators. Many of these plants are also remotely located. Here the key is simple, dependable controls.

Current technology in process control of gas processing units is by distributed control systems (DCS) and sometimes programmable logic controllers (PLC). These units are dedicated control processors. Individual gas processing unit operation control and optimization strategies deployed in these systems are discussed in this chapter.

3.6.1 Blending

Continuous blending is typically accomplished with ratio control. There are two basic control strategies for blending. The first strategy includes controlling a main blend component on flow control and basing the other blend components' flows on ratios to the main component. The second strategy includes blending the individual components based on a ratio of the total blend flow.

When conditions require tight composition control, an analysis will be the master controlled variable that usually resets the ratio(s). For faster reaction to changes in composition of a blend component, the control strategy may include feedforward.

If the main blending components are similar in rate, then there may be interactions that can cause an upset when one blending component is manipulated. Elimination of the interaction requires decoupling methods mentioned in Section 3.3.6.4.

Optimization typically includes minimization of the more expensive components while meeting product quality or minimization of an additive for an intermediate stream. Mercaptan injection into a sales gas stream is one example.

3.6.2 Boilers

Boilers are often employed in natural gas processing operations to supply process heat and to drive steam turbines. There are many varieties of boilers including fire tube and water tube types as well as those producing saturated steam or superheated steam. Some boilers are forced draft where air is blown into the furnace section, induced draft where flue gas is drawn out of the top of the boiler, or balanced draft with both forced and induced draft. Most boilers in natural gas processing are fueled by natural gas. This section will focus on natural gas—fueled boilers.

The prime control consideration for boilers should be safety assurance. Most boilers must meet minimum safety requirements per National Fire Protection Association codes and frequent inspections by insurance underwriters.

A secondary control consideration is emissions. Many boilers are regulated by federal or state agencies and permitted for maximum emissions of nitrous oxides, sulfur oxides, carbon monoxide, and sometimes carbon dioxide. Most natural gas—fired boilers produce relatively low emissions and flue gas treating is normally not required.

Opportunities for control and optimization occur due to changes in steam demand, fuel, ambient conditions, and boiler condition while meeting safety and emissions constraints. In addition, there may be several boilers operating in parallel and delivering steam into a common header. These situations provide opportunities for load sharing and optimum allocation.

Boilers are typically well instrumented for safety considerations, but do not always have adequate instrumentation for emissions control and optimization. The main instrumentation that is often lacking are flue gas analyzers to measure emissions and oxygen and temperature.

Interlocks and relief valves provide the main safety controls for boilers. Typical interlocks include:

- Purge
- Low fuel supply
- Low air flow or combustion air fan loss
- Flame loss
- High combustibles in fuel gas
- High and low furnace pressure
- Low water level in boiler

Boiler combustion air is normally controlled through a combination of feedforward and feedback methods. Typically steam demand will drive the amount of fuel required and a proportionate amount of air will be introduced. The air flow is often manipulated by dampers, but fan control is preferred. Variable pitch fan blades or a variable speed fan drive uses less energy. Dampers on constant pitch and speed fans dissipate excessive energy input through throttling.

A continuous oxygen composition of the flue gas taken as close to the combustion zone is desired to minimize the effects of air leakage. This continuous oxygen reading is converted to excess air and corrects the air to fuel ratio on a feedback basis.

It is advisable to employ start-up, shutdown, and normal operation control schemes to err on the side of too much air for safety considerations. However, lower excess air leads to higher boiler efficiency and generally lower nitrous oxide (NO_x) emissions. A combination combustibles/carbon monoxide analyzer is recommended to detect a point where air is inadequate and efficiency is lost due to incomplete combustion. In addition, a reducing atmosphere will exist that will lead to excessive boiler tube corrosion (Fig. 3.8).

Balanced draft furnaces should be configured so that either the induced draft or forced draft is manipulated to control excess oxygen or the other is manipulated to maintain a slightly negative pressure in the furnace.

Steam pressure is the primary indication of supply and demand imbalances. When the steam pressure reduces, then additional load on the boilers is required to match supply requirements. Alternatively, an increase in steam pressure indicates that supply is greater than demand and the load on the boilers requires an increase.

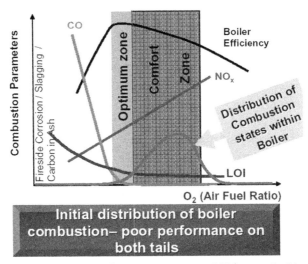

Figure 3.8 Boiler combustion parameters. *LOI*, loss on ignition.

A master controller that monitors steam pressure or other measure of load imbalances typically controls the flow of fuel to a boiler. In the case of natural gas fuel, the supply header pressure is also regulated. Usually, the flow to individual burners is not controlled directly. The pressure to each burner may be controlled with a regulator. Shrouds can also be adjusted to fine-tune burners.

Some boiler control schemes monitor fuel composition and make air–fuel adjustments based on the current fuel composition. Fuel composition variations are common in natural gas processing plants, because the fuel is usually generated internally. During start-up and upset conditions an auxiliary fuel may be used. The control strategy should accommodate a switch to auxiliary fuel with minimal impact on the boiler. The fuel composition can also be used effectively in multivariable control strategies.

Air-limited and fuel-limited conditions are another consideration in fuel to air controls. The maximum capacity of a boiler is commonly the amount of air that can be introduced, which in turn limits the fuel input.

Feedwater and drum level control is quite important in boiler operation. The steam drum receives water treated for reduced scaling and corrosivity. After absorbing heat from the furnace, steam is generated from the treated water. The drum must maintain a level within a controlled range. Too low a level will expose boiler tubes to overheating and leakage, whereas too high a level will lead to insufficient separation between liquid and vapor.

Drum level control is complicated by the fact that steam bubbles entrained in the water contribute to the observed level. This contribution is greater at high loads than at low loads. Therefore the drum level rises or "swells" as the load increases. If the drum level remains constant, then the mass of water in the drum is less at higher loads than at lower loads.

Another drum level control complication occurs because feedwater is colder than the saturated water in the drum. An increase in feedwater flow causes some steam to condense, reduce evaporation in the boiler tubes, and collapse some bubbles in the drum leading to a characteristic called "shrink." The initial reaction to feedwater flow at constant load can actually lead to a decrease in level or an inverse response. After equilibrium is restored, the level will then begin to rise. If a constant level is desired to honor boiler design considerations, then special control strategies are required.

For small boilers with gradual changes in loads and relatively large drum storage volume, feedwater feed alone can be manipulated to control drum level. A large proportional band with no integral action is recommended. This strategy is called single element control.

As boiler size and load variability increases with relatively small drum storage volume, a two-element system is typically used. The two-element system includes a characterization of steam flow to feedwater flow control valve combined with the level signal preferably measured with displacement or differential pressure (dp) cell type sensor.

Three-element systems are used for large boilers designed with small drum storage volumes and high velocities of water and steam. The three-element system replaces the level controller with a flow ratio of steam and feedwater controlled after level and steam flow effects are combined.

Feedforward control is a fourth option for drum level control. In this scheme, the steam to feedwater ratio is compared and used as the process variable for the feedwater flow controller. The level controller resets the set point for the desired ratio between steam and feedwater flows.

In any of the drum level control strategies, feedwater pump speed can be regulated in lieu of a feedwater flow control valve. Knowledge of the pump head curve is necessary for the two-element, three-element, and feedforward strategies.

Steam temperature is a complicated function of heat exchange area, flue gas mass flow and temperature, flue gas flow pattern, and steam flow. Efficiencies of steam turbines and heat exchangers are maximized at the highest steam temperatures tolerable. The tolerance limits are typically metal temperatures for boiler and exchanger tubes as well as turbine blades. Exceeding steam temperature targets is a major factor in reduced boiler reliability.

Flue gas recirculation or bypass is a common method of steam temperature control, which affects fuel gas mass flow and temperature. This control can be supplemented with attemperators. Attemperators are inefficient for steam temperature control and should be minimized. Therefore the flue gas flow and temperature should be the primary control, with attemperator control a secondary method. Flue gas temperature, however, should be maintained above dew point as condensate will contain acids that will lead to corrosion of boiler components.

Burner tilts, when available, and excess air also affect steam temperature. Boiler stack losses versus increased turbine thermal efficiency should be considered when using variable excess air.

NO_x emissions are becoming a key constraint in boiler operations. Boilers in most parts of the world are regulated for a maximum amount of NO_x emissions as well as CO, opacity, and other pollutants measured on an instantaneous, average, or cumulative basis. NO_x formation is primarily a function of excess oxygen and flame temperature. Lower excess oxygen and flame temperatures lead to less NO_x formation. Therefore NO_x reduction strategies should include excess air and staging of air entry through secondary and overfire air introduction, when available.

Optimization opportunities typically focus on energy reduction, but can be capacity driven when steam production limits overall plant production. In today's environment, emissions will constraint how far energy can be reduced or steam production can be increased. Areas of optimization include combustion, steam pressure, blowdown, and load allocation.

Boilers are prime candidates for multivariable control due to the variety of operating constraints that may be imposed. A typical multivariable control scheme will drive the excess oxygen as low as possible for energy reduction and NO_x considerations. However, CO and opacity limits must be honored. This can be accomplished with regulatory control systems that employ selectors. However, a model predictive system is necessary for more complicated systems with many manipulated variables such as secondary air, gas recirculation and overfire air. Flue exit gas temperature can also be monitored and taken into account with multivariable control systems.

Steam pressure optimization or sliding pressure control is another effective strategy that can be included in a multivariable control and optimization strategy.

When the steam generated by a boiler is used to drive steam turbines, the maximum turbine horsepower is achieved at maximum boiler pressure and minimum turbine exhaust pressure. This is effective when electricity is produced by a steam turbine—driven generator and sold or used to reduce purchased electricity. Also, the maximum boiler pressure is desirable when production capacity is restrained by a steam turbine driver.

Alternatively, when all steam users are satisfied, the boiler pressure should be reduced. Otherwise, equipment such as steam turbine—driven pumps will be throttled to control discharge pressure, thereby wasting the additional energy expended in the boiler. This condition can be effectively optimized by monitoring the position of the steam user valves. If all of these valves are less than fully open, then lower steam pressure will not restrict steam availability to the users. Boiler efficiency is also increased as losses due to radiation, heat transfer through the walls, and feedwater pumping are reduced. The overall steam system stability improves and turbine drive efficiencies increase when steam governors are opened.

Water blowdown should be minimized without allowing scale buildup inside the boiler tubes. Heat loss from the discharged water, makeup water, and treating chemicals contribute to the savings from blowdown minimization. Depending on the treatment method, chlorides and conductance are monitored and controlled for proper blowdown. The cycles of concentration of chloride or other key contributor to scaling should be monitored for boiler performance. Cycles of concentration are calculated as:

$$C = (V + B - L)/B \qquad (3.50)$$

where C is cycles of concentration, V is the steam production rate, B is the blowdown rate, and L is the condensate return rate.

Optimal allocation of boiler load allows the most efficient steam production when multiple boilers are employed. Different boilers will have different efficiencies due to fuel, design, mechanical condition, and operating conditions. Optimization can be open-loop advisory or closed-loop supervisory. Closed-loop supervisory optimization is simplified when all equipment runs continuously. The solution requires discrete integer optimization when determining which equipment to run is part of the objective.

In the simplest load allocation systems, the starting and stopping of boilers is optimized recognizing that there is an incremental cost in starting and stopping a boiler as well as a possible discontinuous impact on production. The least expensive idle boiler will be started when load is increasing and the most expensive running boiler will be stopped when load is decreasing.

In more complex load allocation systems, the distribution of load among boilers is constantly monitored adjusted based on load, fuel costs, and overall utility, as well as production considerations. The operating characteristics of each boiler can be modeled and updated with real-time information to fine-tune the optimal solution.

3.6.3 Utility Steam Systems

In gas processing applications, the main purpose of the boilers is the generation of utility steam. Utility steam is typically supplied at several pressure levels for supplying process heat and powering steam turbines.

Often steam is produced at a single, highest-pressure level. The most efficient method of reducing pressure is through steam turbines. Otherwise, the pressure is decreased through pressure letdown valves to provide lower pressure steam. Although the letdown valves operate isenthalpically, steam turbines are more effective due to an approach to an isentropic process of a steam turbine driving a compressor, pump, or generator. Also, the letdown steam often requires desuperheating.

Obviously low-pressure steam venting is not desirable, but is sometimes necessary due to supply and demand imbalances. In a few instances when additional power can be produced and consumed or sold at attractive pricing, the steam venting can be economically viable.

3.6.4 Steam Turbines

A steam turbine is a mechanical device that extracts thermal energy from pressurized steam, and converts it into useful mechanical work. It has almost completely replaced the reciprocating piston steam engine, primarily because of its greater thermal efficiency and higher power to weight ratio. Also, because the turbine generates rotary motion, rather than requiring a linkage mechanism to convert reciprocating to rotary motion, it is particularly suited for driving an electrical generator—about 86% of all electric generation in the world is by use of steam turbines.

Steam turbines are configured to use steam in several configurations including:
- Condensing
- Noncondensing

- Extraction
- Topping
- Bottoming

Steam turbines may also have multiple stages.

Condensing steam turbines operate with a subatmospheric pressure exhaust. Noncondensing or back pressure turbines operate with the exhaust pressure greater than atmospheric pressure. Steam turbines may have one or more outlets in addition to an exhaust that provides steam at intermediate pressure for other steam system uses.

Turbines that take inlet from the high-pressure header and exhaust steam to an intermediate pressure are called topping turbines. The exhaust of another turbine supplies a bottoming turbine while the discharge is condensate.

Steam turbines are inherently variable in speed. With modifications to the governor, they can be operated at a wide variety of speeds. The efficiency of steam turbines is maximum as superheat temperature increases and with the amount of vacuum pulled on the turbine exhaust.

Safety systems are quite important for steam turbines. Typical elements of the safety system are:

- Steam supply valve
- Overspeed
- Lube oil
- Bearing temperature

The steam supply valve is the main element of the safety system and shuts down the equipment when closed. Overspeed is normally caused by loss of load and detected by a tachometer, switches, or strain gauges. Governors are normally employed to limit turbine speed and regulate the opening of the steam supply valve.

Lube oil is generally monitored by pressure or level. High bearing temperature is another method of detecting lube oil failure, but it also may be an indication of deformed bearings.

Machine health monitoring or damage has conventionally been detected by vibration as measured by accelerometers or proximity sensors. Stress wave analysis has been employed to monitor equipment through analysis of ultrasonic signals.

In addition to the steam supply valve that regulates the input of steam to the turbine, extraction turbines include a second valve or assembly inside the turbine to control the rate of steam extraction. Typically, the outlet pressure controls the rate of extraction, but speed or a combination of speed and pressure can be used.

Low-pressure steam is preferentially supplied by extraction or topping turbines. However, when the rate of extraction is not adequate to satisfy the low-pressure header as detected by header pressure, then high-pressure steam will be let down to supplement low-pressure supply. Pressure letdown leads to lower overall efficiency of the utility steam system. In addition, if the temperature of superheated steam exceeds the allowable metal temperatures, then the low-pressure steam will require desuperheating.

One method to maximize the overall efficiency of a utility steam system is to install a flow controller on the letdown line. When the flow exceeds a high limit (set as low as possible), then the turbine speed is increased. This is most effective when the turbine load can be increased such as cogenerated electricity can be increased. Low header pressure should remain under the control of the letdown valve, since it will react faster than the turbine to undesirable pressure conditions.

Consider a two-stage extraction turbine as shown in Fig. 3.9. When the low- or intermediate-pressure header does not need all the steam provided by the extraction turbine and the letdown valve is closed, then load should be shifted from the first stage to the second stage.

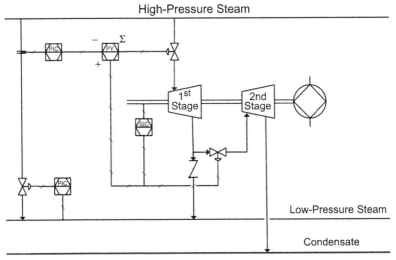

Figure 3.9 Two-stage extraction turbine. *PIC,* Pressure indicating controller; *PY,* pressure relay (standard control system symbology); *SIC,* speed indicating controller.

The control of a turbine with a governor is essential, as turbines need to be run up slowly to prevent damage. Some applications, such as the generation of electricity, require precise speed control. Uncontrolled acceleration of the turbine rotor can lead to an overspeed trip, which causes the nozzle valves that control the flow of steam to the turbine to close. If this fails, the turbine may continue accelerating until it breaks apart, often spectacularly.

The main functions of a modern steam turbine control system are:
- Speed and acceleration control during start-up
- Initialization of generator excitation
- Synchronization and application of load in response to local or area generation dispatch commands
- Pressure control of various forms: inlet, extraction, back pressure, etc.
- Unloading and securing of the turbine
- Sequencing of the above-mentioned functions under constraint of thermal stress
- Overspeed protection during load rejection and emergencies
- Protection against serious hazards, e.g., loss of lube oil pressure, high exhaust temperature, high bearing vibration
- Testing of steam valves and other important protective functions.

Additional control and monitoring functions are also required in most applications, such as:
- Monitoring and supervision of a large number of pressures, temperatures, etc., to provide guidance and alarms for operators
- Start-up and monitoring of turbine-generator auxiliaries such as lube oil, hydraulic, and steam seal systems
- Display, alarm, and recording of the above-mentioned functions and data
- Diagnosis of turbine or generator problems
- Health check and diagnostics of the electronic system itself

3.6.5 Heat Exchangers

Heat exchangers are generally controlled by either throttling the heating fluid (e.g., steam or heat-transfer fluid) or by bypass control of the process fluid. Flow control of cooling water is not recommended, even though it would work as a temperature control loop. Reduced cooling water velocities can increase fouling. Scaling and fouling also increase dramatically when cooling water temperature exceeds 120°F. In an extreme case, vaporization of the water can occur.

A common bypass control scheme uses a three-way valve.

3.6.6 Chillers

A chiller is a machine that removes heat from a liquid via a vapor-compression or absorption refrigeration cycle. A vapor-compression chiller uses a refrigerant internally as its working fluid. Many refrigerant options are available; when selecting a chiller, the application, cooling temperature requirements, and refrigerant's cooling characteristics need to match. Important parameters to consider are the operating temperatures and pressures. Natural gas processing employs a number of refrigerants. Among some of the refrigerants used are:

- sodium brine
- ammonia
- freon (phased out)
- propane
- ethylene (often cascaded with propane), and
- mixed light hydrocarbon refrigerants.

In some processing schemes, the refrigerant is delivered directly to chillers and in other schemes chilled water or other medium is delivered to the process.

The typical refrigeration cycle consists of two isothermal and two isenthalpic or adiabatic steps at isobaric conditions. Usually, the heat of vaporization isothermally transfers heat from the refrigeration to the process yielding a lower temperature of the process fluid. The refrigerant vapors are compressed to raise the pressure adequately for condensing by a heat sink. The heat sink is typically cooling water or ambient air; however, in cascade refrigeration schemes the heat sink may be a higher–temperature refrigerant.

The heat sink is of sufficiently low temperature to condense the refrigerant totally and isothermally at the discharge pressure of the refrigerant compressor. Condensed refrigerant, now in liquid form, is expanded isenthalpically across a valve to lower the temperature of the liquid refrigerant. A portion of the refrigerant will flash into vapor across the valve. This vapor is dead load in the chiller unless the refrigerant pressure is lowered in stages with economizers installed to separate the liquid from the vapor. The remaining liquid will enter the chiller to continue the refrigeration cycle (Fig. 3.10).

Point A to B is the isenthalpic expansion step.
Point B to C is vaporization of the refrigerant.
Point C to D is the compression step.
Point D to A is the refrigerant condensation step.

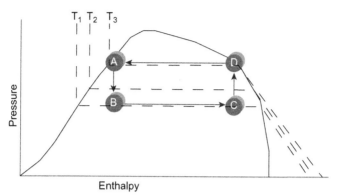

Figure 3.10 Refrigeration cycle.

In most systems, the refrigerant is condensed by water from a cooling tower or air from fin fan coolers. The main cost of condensing in either case is the horsepower required for circulating the air and cooling towers require the additional costs of pumping and treating the cooling water.

In general, when adequate refrigeration is available, the highest refrigerant supply temperature that will satisfy the process needs will result in the least cost. Often lower process temperatures will result in higher recovery, yield, or throughput and then the lowest refrigerant temperature (lowest refrigerant compressor suction pressure) will result in optimum operation. In all cases, the maximization of the value of incremental recovery or yield less the additional cost of refrigeration will determine the optimum process temperatures. The cost of refrigeration consists of fan motors, cooling tower pumps, and refrigerant compressors.

There will be an optimum approach above the wet bulb temperature for supplying cooling water for refrigerant condensation and an optimum range, which is defined as the difference between the cooling water return and supply. The optimum will be affected by the weather conditions of ambient temperature and humidity that affect wet bulb temperature. Air cooling is similar in that the cost of cooling increases as the temperature rises. The cooling tower water supply temperature should not be held constant. Each 1°F reduction in cooling tower supply temperature results in about a 1.5% improvement in efficiency.

As the approach to wet bulb temperature decreases, the costs of pumping cooling water and compressing refrigerant decrease, whereas the cost of cooling tower fans increases. When less cooling tower fan horsepower is required, fans can be stopped, slowed down, or the blade pitch varied, if these capabilities exist.

The water flows to individual cells of the cooling tower should be adjusted to correspond with fan operation. For instance, cells with fans operating at high speeds should attain high water rates, cells at low speed take low water rates, and cells with fans that are off should receive minimum water rates.

Lower suction pressures normally lead to colder refrigerant temperatures leading to more efficient chilling. However, this creates an increase in the compression ratio for a given condensing temperature. The suction pressure can be optimized by balancing the effect of compression ratio on horsepower (cost) against the increased revenues obtained with colder process temperature.

It is preferential to control evaporator pressure and maintain constant chiller level to fully utilize heat transfer area. Refrigerant levels in chillers should be held adequately high to cover all tubes, but not so high that refrigerant liquid carries over into downstream scrubbers.

Some refrigeration system optimization opportunities may include:
- Maximize temperature difference across the chiller when cooling water is used
- Minimize pumping
- Cooling tower approach optimization
- Store refrigerant at night
- Always use the most efficient chiller combination
- Use efficiency information to initiate maintenance

Ideally, variable speed drives are preferred for refrigeration compressors to match horsepower consumed to amount of refrigerant required. If several constant speed motors are used, then one compressor should be throttled and the remaining compressors should be operated with the throttle valve wide open.

3.6.7 Compressors

The natural gas processing industry employs two primary forms of compressors in variety of services, reciprocating and centrifugal compressors. Rotary compressors are also found in blower applications and other small horsepower purposes. However, rotary compressor control is not covered is this work.

3.6.7.1 Reciprocating Compressors

Reciprocating compressors preferably perform constant volume, variable discharge pressure service. Such compressor controls accomplish three goals:

(1) control compressor capacity, engine load, and speed, (2) control auxiliary items on the packaged compressor, and (3) safety shutdown controls in case of harmful temperatures, pressures, speed, vibration, engine load, and liquid levels (Gilleland, 1979).

Compressor capacity is controlled by varying driver speed, by opening or closing fixed- or variable-volume clearance pockets, or activating pneumatic suction-valve unloaders. Driver speed control is not always provided with synchronous AC motor drivers, although solid-state devices are available for varying speed or preferably input frequency.

On—off control is an effective method of capacity control where suction pressure is monitored. The motor turns on at high suction pressure and the motor is turned off at low suction pressure. This capacity control method is most effective when the continuous usage is less than 50% of maximum capacity.

Unloading is an effective method for saving horsepower on recipro-cating compressors that operate at constant speed. Suction valves are held open on the discharge stroke or clearance pockets in the cylinder are opened. Capacity is incremental to the number of unloading steps. Consider the case of two suction valves. If one suction valve is held open, then the machine is operating at 50% capacity and if both suction valves are held open, then the machine has no capacity. A clearance pocket is utilized to provide 75% capacity when both suction valves are closed and 25% capacity when one suction valve is closed. The unloading steps can be automated with the use of a pressure controller and pressure switches or can be adjusted manually. When several compressors in similar service are available, then lead/lag strategies can be employed to meet demand. Additional compressors can be turned on and loaded when the lead compressor is fully utilized and demand is not met. Time delays are incorporated into the system to prevent too frequent cycling of the lag machines. For compressors that are not identical, the lead machine should be the one with the greatest difference in efficiency between loaded and unloaded conditions, whereas the lag machine should have a greater effi-ciency at low loads.

Typical auxiliary controls include primary and secondary regulators for fuel and starting-gas pressures, engine and compressor jacket-water systems, liquid-level controls for suction and interstage scrubbers, and governors for gas-engine speed control.

The safety shutdown controls ground the engine ignition, close the engine fuel valves, or both. Low lube-oil pressure, high engine jacket-water temperature, high liquid level in scrubbers, high compressor-cylinder cooling-water temperatures, and high discharge temperature or pressure should trigger a safety shutdown.

3.6.7.2 Centrifugal Compressors

Centrifugal compressors come in the axial and centrifugal types. The axial compressor has better characteristics for more constant flow and variable pressure applications, whereas centrifugal configurations are better suited for constant pressure, highly variable flow applications. Poor selection may lead to inefficiencies due to throttling or recycling flow.

Centrifugal compressor control must accomplish three goals, namely, capacity control, surge prevention, and equipment protection.

3.6.7.2.1 Capacity Control Options

Capacity control can be achieved by the following means:
- Speed
- Suction throttling
- Discharge throttling
- Inlet guide vanes

Centrifugal compressors convert momentum into pressure head by accelerating the gas. The ratio between the discharge and the suction pressure varies proportionately with the mass flow and the torque developed by the driver. Suction temperature, suction pressure, and molecular weight will affect the discharge pressure of these compressors at constant speed. Efficiency curves are developed to show this relationship as per Fig. 3.11.

Axial designs that move the gas parallel to the shaft are more efficient for constant flow applications. Centrifugal designs, which provide a radial thrust, are more efficient for constant pressure applications.

Modulating the speed of a centrifugal machine is the most efficient method of turn down due to horsepower savings derived by the cubic relationship of power to speed. Inlet guide vanes are the second best choice as the gas is counterrotated relative to the impeller before compressor entry. Suction valve throttling is the next best choice from an efficiency standpoint, but consumes more horsepower than inlet guide vanes. Discharge valve throttling is the least desirable choice as high energy levels are dissipated across a valve wasting some of the horsepower developed by the compressor. Inlet guide vanes are more complex and expensive than

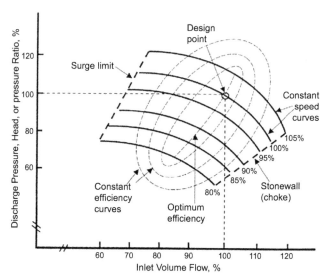

Figure 3.11 Compressor performance map.

throttling valves, but can yield high power savings for applications with wide flow variations.

3.6.7.2.2 Surge

For every compressor speed and inlet gas conditions, a maximum discharge pressure is developed as flow drops. These points create a surge line. Flow reversal occurs at this point and can destroy compressor parts. Operation at flows leading to surge conditions that cause extensive damage must be avoided. Surge is characterized by rapid flow reversals, excessive radial vibration and axial thrust displacement, large temperature rise in the compressor, and booming noise. Flow surges have extremely high frequencies of 50–100 cycles per second and are usually faster than the flow control loop.

Surge protection starts with the determination of the surge limit line (SLL) that is the limiting values of head versus throughput that initiate surge. Surge control keeps the compressor from crossing a surge control line that is artificially set a "safe distance" from the SLL. Protection against surge (antisurge) is typically provided by recycling gas from a discharge stage into the suction of the compressor to increase the flow. Antisurge control requires fast and reliable flow measurement. The flow transmitter must also be placed where noise in the flow measurement is minimized. Extremely fast action is required from surge valves. Linear globe valves with special actuators are typically used.

To control surge, a surge control line is determined that is close to the surge line. The surge control line should be as close to the actual surge line as possible to minimize the inefficiencies of recycling, but far enough that the antisurge controls can act fast enough to prevent surge conditions.

A compressor's surge limit is not fixed with respect to any one measurable variable, such as compression ratio or the pressure drop across a flow meter. Instead, it is a complex function that also depends on gas composition, suction temperature and pressure, rotational speed, and guide vane angle. Algorithms have been developed to adapt the surge curve. Anticipator algorithms have also been used to open the surge valve sooner than needed before low flow conditions are experienced.

A typical set of algorithms may include the following. For small disturbances, PI control is used with provisions for preventing reset windup when the valve is fully closed. Fast disturbances elicit a derivative response that increases the safety margin, thus accelerating the PI control response. If that combination proves insufficient, an open-loop response step opens the antisurge valve even further, using step sizes based on the instantaneous rate of approach to surge. This provides just enough added flow to prevent surge without unnecessary process disruption. Finally, if unanticipated circumstances do produce a surge, a response redefines the surge limit to stop that surge after a single cycle, then remains in effect to prevent future surges.

When the antisurge valve opens, it provides hot recycle gas to the suction side of the compressor. A quench line provides cool gas to the inlet stream via a quench valve to reduce the undesirable temperature rise.

In addition to the SLL, the compressor manufacturer often recommends a high-flow limit on the right-hand side of the operating map shown in Fig. 3.12. Excessive flow across the compressor creates a condition called stonewall, which causes vibration and fatigue failure that may damage the compressor.

In the case of a compression service delivering gas to several users, such as a fuel gas system, the discharge pressure can be reduced to a minimum by monitoring the valve position of the users. A valve position controller can force the maximum opening to 90% by reducing the pressure. This minimizes the pressure drop in the circuit while assuring stable operation and room for response to upsets. The compressor can be slowed down or unloaded to meet the valve position controller requirements. This strategy is referred to as load following.

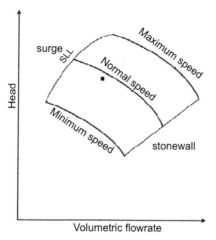

Figure 3.12 Compressor head versus flow map.

When centrifugal compressors are operated in parallel, the problem of properly sharing the load is usually present. The solution is particularly challenging when the compressors have dissimilar characteristics. But even when the characteristics are supposedly identical, load sharing should be provided because minor manufacturing differences and piping configuration affect the operating envelope. An effective system for load sharing can keep compressors at a relative distance from the surge control line but load sharing is normally activated only once all compressors have reached the load-sharing enable line. This load sharing strategy assures that all compressors reach their surge control lines simultaneously. The concept produces the widest range of operation with no recycle (or blowoff) and the minimum recycle/blowoff to produce any required load. In addition, the strategy can optimally divide the load whenever the load changes, when a compressor goes off or on line, or when efficiency changes.

Conventional load sharing strategies employing pressure-to-flow cascade or biasing the output of a single pressure controller are usually ineffective. Base loading, another common method, is also not energy efficient.

Load sharing may be accomplished by the manipulation of a suction valve (as shown in the illustration), by speed changes (if the compressor is turbine driven), or by guide-vane positioning. Thus there are many possible configurations for simultaneous approach to the surge control line.

Since compressors may differ in capacity and characteristics, the absolute distance of the operating point from the surge control line can be meaningful. For compressors with almost identical characteristics, it is recommended that the compressors are allowed to operate unrestricted until the operating point reaches the load-sharing enable line, a line usually set 10% from the surge control line. For compressors with dissimilar characteristics, it is usually preferable to set the load share enable line near the normal operation points, or in other words, the parallel machines operate at equidistance from the surge control line over the complete operation range.

Each compressor controller's load share algorithm engages once the compressors operating have crossed the enable line and holds the throughput/capacity control output (suction valve, turbine speed, or guide-vane) in its position until all load-sharing compressors have reached the enable line. Communication between compressor controllers is via a high-speed digital peer-to-peer link. After all compressors have crossed the load-share enable line, they are kept equidistance from their surge limits, thus helping to minimize recycle or blowoff costs. The equal distribution of the load within the active region (below the load-share enable line) is accomplished by calculating the average distance from the surge line of all compressors on load sharing and then biasing each compressor in the proper direction (Fig. 3.13).

3.6.7.2.2.1 Compressor Unit Safeties Safety devices typically specified for compressors are detailed in (American Petroleum Institute Recommended Practice) API RP 14C. Compressor suction scrubbers are pressure vessels, but they are a part of the compressor system and will not have any pressure control valve or pressure safety valve. However, liquid-level control must be provided.

3.6.8 Turboexpanders

An expander is used to refrigerate a gas stream, by removing work from the gas in a nearly isentropic process. Energy from a high-pressure gas entering the expander is removed and the gas exits at a lower pressure and temperature. The gas flows radially inward and is accelerated through inlet guide vanes and turns an impeller. Work is extracted from the gas by removing the momentum reducing the tangential velocity.

The purpose of the inlet guide vane system is to direct gas flow toward the expander wheel at the correct angle while also providing control of

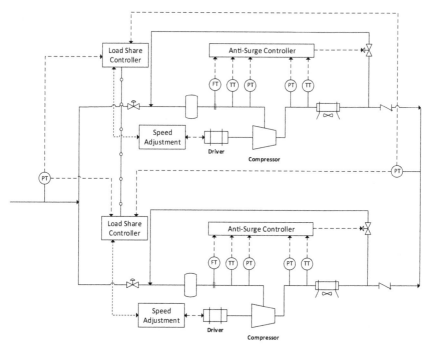

Figure 3.13 Compressor load sharing and antisurge control strategy.

mass flow through the turbine and thereby the rotating speed. Pressure, temperatures, gas composition, material, and clamping forces are all considered to eliminate blow-by and allow the turbine to operate reliably for extended periods. To accommodate varying operating conditions, the inlet guide vane system is designed to open beyond the design point to allow for start-up or emergency situations.

The mechanical power produced by the expander drives a compressor impeller either coupled to the same shaft as the turboexpander or driven via a gearbox. This compressor typically boosts the pressure of the demethanizer overhead vapors. Typically, additional compression is required to meet sales gas pipeline pressure requirements. Demethanizer overhead pressure is usually controlled by the speed of the expander and therefore, the power available to the compressor for boosting the pressure of the gas. Recycle from the compressor discharge to the compressor suction is provided for antisurge control.

A Joule—Thomson valve is installed in parallel to the expander to prevent overspeed conditions.

3.6.9 Distillation

The natural gas processing industry uses distillation in a variety of forms extensively. In general, a distillation system consists of a fractionation tower, condenser, reboiler, and any associated interheaters or intercoolers. Some types of distillation and example applications follow:

- Distillation with condensing and reboiling—(natural gas liquids) NGL fractionators
- Absorbers—acid gas treaters
- Strippers—stabilizers, demethanizers
- Rectified absorbers—lean oil absorbers
- Azeotropic distillation—carbon dioxide recovery
  Other variations for distillation units include:
- Partial condensers—vapor product with reflux
- Total condenser—liquid product with reflux
- Multiple feeds including a split, subcooled feed to demethanizers
- Side reboilers

The operating objective of distillation is product qualities at the top and the bottom of the towers. In some cases, such as absorbers and strippers, there are not enough degrees of freedom to control the product quality at the top and the bottom simultaneously. In a few cases in natural gas processing, there may be side draws, which may be controlled as well.

Product quality specifications are the primary objective of most distillation operations. Minimizing the energy consumption or maximizing the throughput while continuing to meet minimum product quality is a secondary objective. Throughput maximization or energy minimization is achieved when one or more constraints are encountered. The ultimate constraint is product composition specification, but other operating constraints may provide a warning of off-specification product including:

- High bottom level indicating inadequate low suction pressure for a product bottom pump, inadequate pump horsepower, or too low a differential to flow into downstream equipment
- Maximum heat input to the column indicated by a heat medium flow valve that is wide open
- Maximum cooling to the column indicated by a cooling medium flow valve that is wide open
- High tower differential pressure indicating flooding of the distillation column

• High condenser level indicating low suction pressure for a product bottom pump, inadequate pump horsepower or too low a differential to flow into downstream equipment
• High tower pressure indicating inadequate condensing, lack of compression horsepower for a partial condenser or too low a differential to flow into downstream equipment or flare

3.6.9.1 Steady State Modeling—Material Balance, Energy Balance

To control a distillation column, the process model must be understood. Material and energy balances are required to model the process.

An assumption of equimolal overflow, which is conditional on equal heats of vaporization for all feed components, negligible heat of mixing and no heat loss, keeps the model relatively simple.

3.6.9.2 Dynamic Model

Distillation columns are very difficult to control largely because their response characteristics are inherently nonlinear. In addition, the liquid holdup on the trays causes inherent time delays or lags, and, therefore, a "sluggish" overall response.

3.6.9.3 Control Loop Interaction

In addition to an understanding of the process dynamics, a relative gain array analysis will reveal the control scheme configuration that will give minimum interaction. Some general rules for controlled and manipulated variable pairings are given in the following table:

	Manipulated variable			
	Distillate flow	Bottoms product flow	Reflux flow rate	Vaporization rate
Controlled variable				
Overhead product purity	Yes, if reflux ratio is less than about six	No	Yes, subject to energy and material balance	Yes, subject to energy and material balance
Bottom product purity	No	Yes, subject to energy and material balance No	Yes, if less than 20 trays	Yes, subject to energy and material balance

	Manipulated variable			
	Distillate flow	Bottoms product flow	Reflux flow rate	Vaporization rate
Reflux accumulator level	Yes, if reflux ratio is greater than about six		Yes, if reflux ratio is greater than 1/2	Yes, if vaporization to bottoms product flow is less than three and no furnace
Bottoms level	No	Yes, if vaporization to bottoms product flow is less than three	No	Yes, if bottom diameter is less than 20 feet and no furnace

3.6.9.4 Composition Control

Product quality is theoretically determined by the heat balance of the distillation column. Heat removal determines the internal reflux flow rate, whereas heat addition determines the internal vapor rate. These determine the circulation rate, which determines separation between two components. The first task in configuring a column's control system is to configure the primary composition control loops. The most pure product should be controlled by the energy balance.

3.6.9.4.1 Composition Control Using Analyzers

Analyzer controllers in a feedback configuration should only be considered when the dead time caused by analysis update is less than the response time of the process (Liptak, 1995). The composition controller provides a feedback correction in response to feed composition changes, pressure variations, or changes in tower efficiencies. In Fig. 3.14, the analyzer controller uses the chromatographic measurement to manipulate the reflux flow by adjusting the set point to the reflux flow controller.

The purpose of cascade systems is to provide secondary or slave controllers that will correct for disturbances before they can upset the primary or master controller. Fig. 3.15 illustrates a triple cascade loop, where a temperature controller is the slave of an analyzer controller, whereas the reflux flow is cascaded to temperature. The slave must be faster than the master, which adjusts its set point. Therefore the time constants of the flow controller must be much smaller than those of the temperature recording controller.

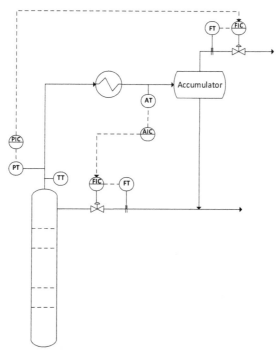

Figure 3.14 Composition control by cascading reflux flow as slave controller. *ARC*, analysis recording controller; *AT*, analysis transmitter; *FRC*, flow recording controller; *FT*, flow transmitter; *TT*, temperature transmitter; *PRC*, pressure recording controller; *PT*, pressure transmitter; *TRC*, temperature recording controller.

3.6.9.4.2 Composition Inferentials—Pressure-Compensated Temperature

Temperature measurements can often be used to infer composition. This is very desirable when an online analyzer is not feasible or an estimate of the composition is needed during periods when the analyzer is not available due to failure or maintenance. Composition inferentials also maintain product quality in between analyzer updates.

Distillation works on the principle of separation of components by their differences in vapor pressure. Vapor pressure is a function of temperature, pressure, and composition. Since the column pressure may vary a pressure-compensated temperature is preferred. Determine the best tray locations for inferential control by finding the trays whose temperatures show the strongest correlation with product composition.

For most systems, a simple linear correction compensates for variations in column pressure:

$$T_{pc} = T_{meas} - K_{pr}(P - P_o) \tag{3.51}$$

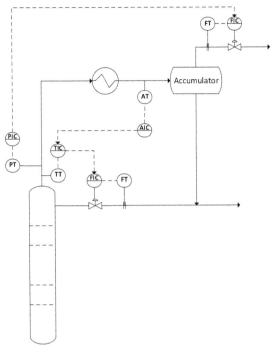

Figure 3.15 Control by triple cascade of analyzer controller to TRC to flow controller. *ARC*, analysis recording controller; *AT*, analysis transmitter; *FRC*, flow recording controller; *FT*, flow transmitter; *TT*, temperature transmitter; *PRC*, pressure recording controller; *PT*, pressure transmitter; *TRC*, temperature recording controller.

where T_{pc} is the pressure-compensated temperature that should be used for feedback control, T_{meas} is the measured tray temperature, K_{pr} is the pressure correction factor, P is the operating pressure, and P_o is the reference pressure.

K_{pr} can be estimated by applying a steady-state column simulator for two different pressures within the normal operating range and using the following equation:

$$K_{pr} = (T(P_1) - T(P_2))/P_1 - P_2 \qquad (3.52)$$

where T is the value of the tray temperature predicted by the column simulator.

For multicomponent separations, tray temperatures (even pressure compensated) do not uniquely determine the product composition. For these cases it is essential to use an online analyzer or periodic laboratory analysis to adjust the tray temperature set point to the proper level.

3.6.9.5 Pressure Control

Floating pressure control whereby the minimum pressure is achieved realizes several advantages for distillation column operation. These include:
- Reduced heat input to effect a certain separation
- Greater reboiler capacity
- Reduced reboiler fouling

Typical constraints that must be monitored when minimizing tower pressure include:
- Bottom tower pressure to satisfy head for pumping or feeding a downstream tower
- Tower differential pressure, which indicates flooding
- Condenser capacity to make reflux

3.6.9.6 Liquid Distillate and Inerts

The presence of large quantities of inerts or noncondensables complicates pressure control. If not bled, the inerts will build up and render the condenser surface unusable. Since a liquid distillate system is designed to condense reflux and product, a high point vent is recommended to control pressure when the primary pressure controller passes a certain operating point.

If inerts are negligible and the condenser has adequate surface area, then pressure control can be accomplished by directly manipulating the cooling medium. Otherwise, the pressure control valve should be placed between the condenser and the accumulator.

If the condenser is installed below the accumulator, then a hot gas bypass configuration is required where the cooling medium flow is maintained at a constant rate. To accomplish faster pressure response, the vapor rate to the accumulator can be throttled in tandem with hot gas bypass control with the pressure set point of the valve throttling vapor to the accumulator at a higher value than the set point for the hot gas bypass.

The most common configuration that can handle some inerts includes hot gas bypass on pressure control with secondary pressure control on a vent line directly off the accumulator. This configuration provides faster response than throttling liquid from the condenser for pressure control.

3.6.9.7 Vapor Distillate and Inerts

For vapor distillate systems, pressure control is direct as long as the condenser is adequately sized to generate adequate reflux as called for by the accumulator level control manipulating cooling medium flow. If the condenser is undersized, then the hot gas bypass configuration is required, accumulator level is held by the vapor distillate rate, and the hot gas bypass rate around the condenser to the accumulator is manipulated by the pressure controller.

3.6.9.8 Vapor Recompression

Vapor recompression is the evaporation method by which a compressor is used to increase the temperature of the vapor produced. In this way, the vapor can serve as the heating medium for the tower bottom liquid. The efficiency and feasibility of this process lie on the efficiency of the compression and the heat transfer coefficient attained in the heat exchanger contacting the condensing vapor and the boiling liquid. Theoretically, if the resulting condensate is subcooled, this process could allow full recovery of the latent heat of vaporization, which would otherwise be lost if the vapor, rather than the condensate, was the final product; therefore, this method of evaporation is very energy efficient. The evaporation process may be solely driven by the mechanical work provided by the compression.

Pressure control can be achieved by manipulation of the compressor speed. Some cryogenic demethanizers operate on the principal of vapor recompression.

3.6.9.8.1 Feed Control

Control of feed rate and temperature is an important aspect of stabilizing the operation of a distillation unit. A flow controller should maintain as constant a feed rate as possible subject to upstream pressure or level limitations. It is advisable to allow the upstream pressure or level to vary within upper and lower limits. If a steam turbine—driven feed pump is employed, then the flow controller should modulate steam rate.

If feed flow rate, temperature, or composition changes are detected, then these effects should be feedforwarded to the tower.

3.6.9.9 Advanced Controls

Additional control benefits can be derived through control strategies beyond the single input/single output (SISO) control loops for control of product qualities, feed rate, and temperature and tower pressure. These SISO controls require operator intervention to adjust set points to achieve stable performance. Advanced control strategies can go beyond stability and approach optimal performance including the following benefits:

- Increased throughput
- Increased product recovery
- Energy conservation
- Reduced disturbances to downstream units
- Minimize off-specification product
- Reduced operator intervention

Basic SISO controls are designed to keep the unit running, whereas advanced controls are designed to operate the unit at maximum profitability. Sometimes, but not always, additional instrumentation is required. Some advanced control strategies can be configured in DCS and PLC systems; however; most model-based strategies require additional functionality and computing power.

3.6.9.9.1 Feedforward Systems

Feedforward control is the simplest form of advanced control for distillation towers; however, it is somewhat limited in benefits. Effective feedforward strategies compensate for feed flow rate, temperature and composition, ambient temperature effects, reflux flow rate, tower pressure, and reboiler heat medium conditions.

3.6.9.9.2 Dual Composition Control

Direct control of both the top and bottom products yields the benefit of minimized energy requirements. However, interaction issues may create problems with this strategy.

One manipulated variable must satisfy the material balance, whereas the other satisfies the energy balance.

Control of distillate (top) composition can be accomplished by manipulating the distillation flow by material balance of the light key component according to the equation:

$$D = F \cdot (z - x)/(y - x)$$

where D = distillate rate; F = feed rate; x = mole fraction of light key component in the bottom product; y = mole fraction of light key component in the overhead product; z = mole fraction of light key component in feed.

The separation is a function of V/F and the energy balance can be satisfied by determining the boil up (V). Since the boil-up rate is controlled by heat input to the reboiler and relates to V by the equation:

$$Q = V \cdot H_V \qquad (3.53)$$

Q can be determined as:

$$Q = kF \cdot \left[(V/F)_{min} + (V/F) \right] \tag{3.54}$$

Therefore D and Q can be determined on a feedforward basis as the feed rate changes.

3.6.9.9.3 Feed Composition Compensation

Feed composition compensation can be accomplished by simplifying the material balance equation to:

$$D = z \cdot F/m \tag{3.55}$$

where m is the output of the overhead analyzer controller.

$$Q = H_v \cdot [m - K \cdot (B/F)] \tag{3.56}$$

$$\text{Since } B/F = (100 - x_h - z_{ll} - z_l)/(100 - x_d - y_l) \tag{3.57}$$

$$Q = H_V \cdot [m - K \cdot (100 - x_h - z_{ll} - z_l)/(100 - x_d - y_l)] \tag{3.58}$$

The above feed rate and composition effects are based on steady state. Feed composition seldom changes drastically, so a steady-state basis is normally sufficient. However, significant feed rate changes can cause undesirable response. Therefore feed flow compensation should pass through a dynamic element that accounts for dead time and a lag response.

3.6.9.9.4 Internal Reflux

Internal reflux is an important concept for determining the external reflux rate. The liquid flow at the top tray of a column is the internal reflux. A steady state energy balance at the top of the tower produces the internal reflux equation:

$$L_i = L \cdot K_2 \cdot [1 + K_1(T_o - T_r)] \tag{3.59}$$

where L = external reflux rate; K_1 = ratio of specific heat to heat of vaporization of the external reflux; K_2 = ratio of heat of vaporization of external reflux to heat of vaporization of internal reflux; T_o = overhead vapor temperature; T_r = external reflux temperature.

3.6.9.9.5 Supervisory Control

Feedforward controls can become very complicated and still may be inadequate due to interactions. To better handle the calculations necessary for adequate distillation control, a supervisory computer is often employed. This system can update the parameters required for sufficient distillation control. Inferential properties can also be used to estimate product and stream properties in between analyzer updates or when an online analyzer is not available.

Precautions for communications status, bumpless transfer, and honoring operating limits are necessary when implementing supervisory computer control.

3.6.9.9.6 Suboptimization

Feed enthalpy and column pressure are suboptimization variables that can be manipulated to increase profitability. Column pressure can be minimized to effect easiest separation, which yields minimum energy consumption or greater yield of the most valuable product for a given throughput. Feed enthalpy can be manipulated to achieve maximum throughput. Valve position controllers can be employed to drive the column to a constraint such as column differential pressure, reboiler heat duty or condenser capacity.

3.6.9.9.7 Feed Maximization

A valve position controller can be cascaded to the feed flow controller to increase throughput until a constraint is encountered. Often the critical constraint will vary over time. In these cases, a constraint selector or a multiple constraint network is implemented.

3.6.9.9.8 Feed Enthalpy Control

Feed enthalpy control is preferred to feed temperature control since feed quality has more impact on the operation of the tower. Feed enthalpy can be calculated by energy balance around the preheat train and feed quality will be kept constant.

Vapor—liquid circulation below the feed tray decreases relative to above the feed tray when the feed enthalpy is increased. Maximum feed preheat is preferred when the reboiler capacity is limiting or feed preheat is less costly than reboiler heat. When differential pressure above the feed tray or condenser capacity is the constraint, feed enthalpy should be decreased.

Nonlinear, predictive, multiinput/multioutput systems are quite valuable for distillation towers. Although the earlier discussion has presented some concepts for stabilizing and improving the profitability of distillation operations, maximum benefits are seldom obtained due to complications and interactions that most always arise. Model predictive methods have been developed and widely adopted for these operations. Model-based, multivariable control systems and neural net controllers, as well as rigorous optimization, are discussed in subsequent chapters of this work.

3.6.10 Fans

Two types of fans are employed in natural gas processing: axial and radial (centrifugal). Typically, fans are used for air at or below atmospheric pressure and requiring discharge pressures of slightly above atmospheric pressure.

Similar to compressors, fans have a sure point where pulsations and unstable operation may occur. Under low load conditions the blade pitch, driver speed, or inlet vanes can be adjusted, if available, otherwise air must be vented to increase the flow.

One form of flow or pressure control is suction throttling. Although this technique introduces additional driver load, it can allow stable operation at lower flow.

Often multiple fans are used in parallel to provide adequate air for process purposes. Minimum cost is achieved by employing the minimum number of fans and minimizing the power usage of the operating fans.

Variable frequency or speed drives provide maximum flexibility and the lowest operating cost for multiple fan configurations. One fan is used as much as possible throughout the load range by varying drive frequency or speed and without throttling. Once a fan reaches maximum frequency or speed demand, the next fan starts. Logic must be in place to reduce the speed of the first fan when the second fan starts to minimize the disturbance.

The logic reverses when the load decreases to the point that one fan can satisfy the requirement. To prevent excessive starts and stops, a time delay is typically introduced.

The optimum discharge pressure from a fan or a bank of fans is found by monitoring each user's damper position. The discharge pressure can be decreased until one damper position is wide open. This strategy minimizes power consumption.

3.6.11 Furnaces

Furnaces are similar to boilers in many control aspects. For our purposes, furnaces are defined as combustion operations that heat fluids other than water, whereas boilers produce steam. The main objective of indirect furnace operations is to provide heat to a process through a heating medium. Efficient and controlled transfer of the heat of combustion of fuels to the heat medium is accomplished while maintaining operation within safety and environmental limits.

Sometimes furnaces are placed in standby, especially when the primary heat source for a process is the waste heat from a combustion turbine. There are also cases when a furnace is paced in direct heating service such as direct fired reboilers where the process fluid is heated directly.

Combustion is a dangerous operation and precautions must be taken to prevent explosive mixtures from developing within the furnace. This condition is most likely during start-up and shutdown situations. Industry, government, and insurance boards have developed codes and practices to improve safety. Fail open and fail closed situations for air operator valves must be carefully considered. Hazard and operability studies are helpful in analyzing these situations.

Fuel gas, fuel oil, and gas are the most common fuels for furnaces. Sometimes, process waste streams will be introduced as supplemental fuels to furnaces as well. Price, thermal efficiency, and emissions are prime considerations in selecting the fuel.

Typically, the exit temperature of the heat medium fluid is monitored to control the heat transferred in the furnace operation. Precautions must be taken to prevent high tube skin temperatures, which can cause coking, polymerization, or decomposition of the heat medium or process fluid. This temperature must typically be estimated by calculations involving the temperatures of the heat input and the heat medium. Infrared can also be used to measure these temperatures manually.

Sometimes it is most efficient to supply heat medium at a couple different temperature levels. An effective means of accomplishing this is to mix a portion of the return heat medium from the lower temperature service header with furnace exit through temperature control. A header with heat medium at furnace exit temperature serves the higher temperature services.

In all cases, the valve openings for all heating services can be monitored and the supply header temperature can be decreased to a temperature where one valve is at a maximum percentage opening (for example, 90%). This saves energy by not raising the heat medium temperature any higher than normal. More flow is required, but head impacts are minimal since minimum restrictions are imposed by the valves controlling the supply to each user.

To completely burn a fuel, the proper amount of oxygen usually introduced through the admittance of air is required. The quantity above the ideal amount of air is the excess air. Each fuel has a practical minimum of excess air and for maximum efficiency the minimum amount should be controlled.

Typically, the excess air requirement is controlled through furnace draft. Measurement of the oxygen in the flue gas is necessary to monitor the amount of excess air. Measurement of CO will help assure that the minimum amount of excess air is introduced, since CO results from incomplete combustion. One common control strategy is for the oxygen controller to adjust the air to fuel ratio with a low selector for CO to increase the air to fuel ratio when necessary.

Gas and oil fuels should be controlled by maintaining the pressure of the fuel supply header. Atomizing steam is required to disperse the fuel adequately when oil is used. This requires a differential pressure controller to assure that the steam header has sufficient pressure to meet atomizing requirements.

REFERENCES

Åström, K.J., 1989. Toward intelligent control. IEEE Control Systems Magazine 9 (3), 60−64.

Åstrom, K.J., Hägglund, T., 1984. Automatic tuning of simple regulators with specifications on phase and amplitude margins. Automatica 20, 645−651.

Blevins, T.L., McMillan, G.K., Wojsznis, W.K., Brown, M.W., 2003. Advanced Control Unleashed − Plant Performance Management for Optimum Benefits. Instrumentation, Automations, and Systems (ISA).

Boudreau, M.A., McMillan, G.K., 2006. New Directions in Bioprocess Modeling and Control.

Cohen, G.H., Coon, G.A., 1953. Theoretical considerations of retarded control. Transactions of the ASME 827−834.

Dahlin, E.B., 1968. Designing and tuning digital controllers. Instrumentation and Control Systems 41 (6), 77.

Fruehauf, P.S., Chien, I.L., Lauritsen, M.D., 1994. Simplified IMC-PID tuning rules. ISA Transactions 33, 43−59.

Gilleland, R.W., February 25-27, 1979. Control and Maintenance of Field Compressors. SPE 7800, Oklahoma City, OK.

I-Lung, C., Fruehauf, P.S., 1990. Consider IMC tuning to improve controller performance. Chemical Engineering Progress 86 (10), 33−41.

Liptak, B.G. (Ed.), 1995. Instrument Engineers Handbook, Process Control, third ed. CRC Press.

McMillan, G., 2005. Good Tuning − a Pocket Guide, second ed. Instrumentation, Automations, and Systems (ISA).

McMillan, G.K., 1991. Tuning and Control Loop Performance, third ed. Instrumentation, Automations, and Systems (ISA).

McMillan, G.K., 2015. Tuning and Control Loop Performance, fourth ed. Momentum Press.

Morari, M., Zafiriou, E., 1989. Robust Process Control. Prentice Hall.

Morari, M., Garcia, C.E., Lee, J.H., Prett, D.M., 1996. Model Predictive Control. Prentice-Hall, Englewood Cliffs, NJ.

Oglesby, M.J., 1996. Achieving benefits using traditional control technologies. Transactions of the Institute of Measurement and Control 1.

Seborg, D.E., Edgar, T.F., Mellichamp, D.A., 2011. Process Dynamics and Control, third ed. John Wiley, NY.

Shinskey, F.G., June 1993. The effect of scan period on digital control loops. InTech 40 (6).

Shinskey, F.G., 1994. Feedback Controllers for the Process Industries. McGraw-Hill.

Shinskey, F.G., 1996. Process Control Systems, fourth ed. McGraw-Hill, NY.

Skogestad, S., 2003. Simple rules for model reduction and PID controller tuning. Journal of Process Control 13.

Ziegler, J.G., Nichols, N.B., 1942. Optimum settings for automatic controllers. Transactions ASME 759−768.

CHAPTER 4

Process Optimization

4.1 INTRODUCTION

Process optimization is one of the best ways for technically adept organizations operating complex processes to achieve best asset utilization and performance.

Virtually all optimization applications endeavor to directly improve plant profitability or return on investment in an immediately quantifiable way. Almost all processes can benefit from real-time optimization (RTO) in some way by increasing yield and throughput, limiting off-specification production, reducing downtime, and lowering energy costs.

RTO is the process of optimizing some aspects of process design and performance taking into account both steady-state and dynamic considerations. Dynamic optimization, in contrast with simulation, allows you to formulate the problem you are trying to solve directly and then use state-of-the-art numerical and optimization solvers to determine the optimal values or time profiles of the optimization variables.

Objective functions can be arbitrarily defined by combining any variables in the simulation model, such as feedstock, utility and product pricing notes, and information. Constraints can likewise be constructed from any combination of model variables, and can be enforced at certain points in the simulation or over the entire time interval being considered.

The optimization variables are any time-varying controls or time-invariant equipment or other parameters that are normally specified during the simulation. Using rigorous optimization techniques applied to underlying steady states of optimization is in equipment design, design of operating procedures, and in quantifying other key capital and operating decisions. Optimization can be used to give precise, optimal values for quantities such as equipment dimensions, control tuning values, and set point trajectories, all determined simultaneously in the same run, if necessary.

Building the underlying model for optimization requires a true rigorous, first principles modeling tool. Such tools usually take an equation-oriented approach, where the underlying relationships are described by equations

Modeling, Control, and Optimization of Natural Gas Processing Plants
ISBN 978-0-12-802961-9
http://dx.doi.org/10.1016/B978-0-12-802961-9.00004-8

representing the process physics and chemistry. The resulting set of equations is solved transparent to the user.

RTO is a type of closed-loop process control that attempts to optimize process performance (usually measured in terms of profit) online, in real time. These closed-loop control systems are distinct from traditional process controllers, in that they are built upon large-scale, model-based optimization systems.

4.2 TYPES OF OPTIMIZATION

Optimization problems can be classified by the following (Yang, 2010):
- Number of objectives (single objective or multiobjective)
- Number of constraints (unconstrained or constrained)
- Type of constraints (equality or inequality)
- Landscape (unimodal or multimodal)
- Function form (linear or nonlinear)
- Variables/Response (continuous or discrete)
- Determinancy (deterministic or stochastic)

Most real-world optimization problems are multiobjective and constrained. Constraints can be equalities or inequalities.

If there is only a single valley or peak with a unique global optimum, then the optimization problem is unimodal. In this case, the local optimum is the only optimum or global optimum. In engineering optimization, we tend to design or reformulate the problem in terms of a quadratric matrix form, which are often convex and the optimum solution is the global optimum. Multimodal functions are much more difficult to solve.

Discrete optimization consists of integer programming or combinatorial optimization. This class of optimization includes vehicle routing and airline scheduling problems. If all the design variables are continuous or real values in some interval, the optimization is called continuous. In some problems, the design variables can be both discrete and continuous. This type of optimization is the mixed type or mixed integer programming.

Optimization can be divided into deterministic and stochastic algorithms. If there is any uncertainty in the design variables, objective function or constraints, then the optimization is not deterministic (known for certain), but stochastic. Sometimes stochastic optimization is called robust optimization with noise. For most stochastic problems, the problem must be redefined for standard optimization techniques to apply. Redefinition may involve the averaging over some space and the objective evaluated in terms of mean or related uncertainties.

Deterministic methods follow a rigorous procedure and are repeatable for the same starting point. Stochastic algorithms have randomness. The solution will be different each time since the algorithms use pseudorandom numbers. Although the final result may not be much different, the path is not repeatable.

Hybrids of deterministic and stochastic algorithms can be used to determine a global optimum where a deterministic algorithm may be limited to a local optimum.

There are also static problems referring to optimization at one instant in time only or dynamic problems, which are solved over a finite or infinite time horizon.

4.3 CONVENTIONAL OPTIMIZATION TECHNIQUES

Most conventional algorithms are deterministic. These techniques include:
- Gradient-based methods
  - Steepest descent method
  - Line search
  - Newton's method
  - Conjugate gradient method
- Linear programming
  - Simplex method
- Broyden—Fletcher—Goldfarb—Shanno (BFGS) method
- Nelder—Mead method
- Trust region method
- Sequential quadratric programming

Gradient-based methods search directions defined by the gradient of the function at the current point.

The line search approach first finds a descent direction along which the objective function will be reduced and then computes a step size that determines how far the next step should move along that direction. The step size can be determined either exactly or inexactly.

Steepest descent is a first-order optimization algorithm. To find a local minimum of a function using gradient descent, one takes steps proportional to the negative of the gradient (or of the approximate gradient) of the function at the current point. If instead one takes steps proportional to the positive of the gradient, one approaches a local maximum of that function; the procedure is then known as gradient ascent.

Newton's method uses curvature information to take a more direct route by using the function values and their derivatives. It works well for smooth unimodal problems.

The conjugate gradient method is often implemented as an iterative algorithm, applicable to sparse systems that are too large to be handled by direct methods. The conjugate gradient method can also be used to solve unconstrained such as energy minimization (Hestenes and Stiefel, 1952).

The simplex method is the most popular method for solving problems in linear programming. This method, invented by George Dantzig in 1947, tests adjacent vertices of the feasible set (which is a polytope) in sequence so that at each new vertex the objective function improves or is unchanged.

The BFGS algorithm is an iterative method for solving unconstrained nonlinear optimization problems (Fletcher, 1987).

The BFGS method approximates Newton's method. For such problems, a necessary condition for optimality is that the gradient be zero. The BFGS method, like Newton's method, uses both the first and second derivatives of the function and is not guaranteed to converge unless the function has a quadratic Taylor expansion near an optimum. However, BFGS has proved to have good performance even for nonsmooth optimizations.

The Nelder—Mead method, also called the downhill simplex or ameba method, is commonly applied to optimize in a multidimensional space. It is applied to nonlinear optimization problems for which derivatives may not be known. However, the Nelder—Mead technique is a heuristic search method that can converge to nonstationary points on problems that can be solved by alternative methods (Nelder and Mead, 1965).

Trust region is a subset of the region of the objective function that is approximated using a model function that is often a quadratic (Sorensen, 1982). If an adequate model of the objective function is found within the trust region, then the region is expanded; conversely, if the approximation is poor then the region is contracted. Trust region methods are also known as restricted step methods.

The fit is evaluated by comparing the ratio of expected improvement from the model approximation with the actual improvement observed in the objective function. Simple thresholding of the ratio is used as the criterion for expansion and contraction. A model function is "trusted" only in the region where it provides a reasonable approximation.

Trust region methods are in some sense quite similar to line search methods: trust region methods first choose a step size (the size of the trust region) and then a step direction, whereas line search methods first choose a step direction and then a step size.

Sequential quadratic programming (SQP) is an iterative method for nonlinear problems. SQP methods are used when the objective function and the constraints are twice continuously differentiable.

4.4 LIMITATIONS OF OPTIMIZATION

Care must be taken to determine that the global optimum is found rather than a local optimum, which would be a global suboptimum. Nonlinearity and multimodality are the main reason for conventional methods finding suboptimal solutions. If there is some discontinuity in the objective function, then a gradient-based algorithm will not work well.

Another challenge can be the number of decision variables. Non-linearities combined with the number of possible solutions can exceed the computing power of most computers rendering the search of all possible combinations impractical. Heuristic and metaheuristic algorithms are designed to deal with this challenge.

All optimization problems are formulated so that the objective and constraint functions are determined exactly. In reality, all measured parameters have some uncertainty. When there is uncertainty and noise, the optimization becomes a stochastic optimization problem, sometimes called robust optimization with noise. The problem must be redefined or reformulated so standard optimization techniques can be used.

4.5 METHODS OF OPTIMIZATION

Optimization methods must consider existence of an optimum. The extreme value theorem of Karl Weierstrass states that a continuous real-valued function on a compact set attains its maximum and minimum value. More generally, a lower semicontinuous function on a compact set attains its minimum; an upper semicontinuous function on a compact set attains its maximum.

For unconstrained problems, one of Fermat's theorems states that optima are found at stationary points, where the first derivative or the gradient of the objective function is zero. More generally, optima may be found at critical points, where the first derivative or gradient of the objective function is zero or is undefined, or on the boundary of the choice set. When the first derivative equals zero the first-order condition is satisfied and defines an interior optimum.

Optima of equality-constrained problems can be found by the Lagrange multiplier method. The optima of problems with equality and/or inequality constraints can be found using Kuhn—Tucker conditions (Kuhn and Tucker, 1951).

Although the first derivative test identifies points that might be extrema, this test does not distinguish a point that is a minimum from one that is a maximum or one that is neither. When the objective function is twice differentiable, these cases can be distinguished by checking the second derivative or the matrix of second derivatives (called the Hessian matrix) in unconstrained problems, or the matrix of second derivatives of the objective function and the constraints called the bordered Hessian matrix in constrained problems.

To distinguish maxima or minima from other stationary points, a second-order condition is tested with the second derivative test. If a candidate solution satisfies the first-order conditions, then satisfaction of the second-order conditions as well is sufficient to establish at least local optimality.

The candidate solution is a local maximum when the second derivative is less than zero, a local minimum when the second derivative is greater than zero, and inconclusive when the second derivative equals zero. In the case of the second derivative equal to zero, Taylor's theorem may be used to determine the behavior using higher derivatives.

Many optimization algorithms need to start from a feasible point. One way to obtain such a point is to relax the feasibility conditions using a slack variable. This slack variable will turn an inequality constraint into an equality constraint. Given enough slack any starting point is feasible, then minimize that slack variable until slack is null or negative.

The method of optimization chosen must meet the type of problem to solve. The main characteristics of a problem to discern before choosing a method are:

- Continuous, discrete, or a mixture of continuous and discrete
- Linear or nonlinear
- Stochastic or deterministic
- Steady state or dynamic

Formulations of the various types of problems and some typical solutions are discussed in the following subsections.

4.5.1 Continuous Optimization

Continuous optimization techniques exploit mathematical programming formulations (Jeter, 1986). For the basic mathematic programming problem, f is a function of variables $x_1, x_2, \ldots x_n$ where the objective is to:

$$\text{minimize or maximize } f(x_1, x_2, \ldots x_n) \qquad (4.1)$$

subject to

$$(x_1, x_2, \ldots x_n) \vee W$$

where W is a subset of the domain of f. If $W = \{(x_1, x_2, \ldots x_n): \text{each } x_i \vee \mathbf{R}\}$ where $\mathbf{R}$ denotes the set of real numbers, then Eq. (4.1) is said to be unconstrained. Eq. (4.1) is constrained whenever W is a proper subset of $\{(x_1, x_2, \ldots x_n): \text{each } x_i \vee \mathbf{R}\}$. When Eq. (4.1) is constrained, the set W is usually defined by a system of equations and inequalities that are called constraints. The function $f(x_1, x_2, \ldots x_n)$ is called the objective function, whereas W is the set of feasible solutions.

4.5.2 Linear

The function $f(x_1, x_2, \ldots x_n)$ is linear if it has the form:

$$f(x_1, x_2, \ldots x_n) - c_1 x_1 + \ldots + c_n x_n \qquad (4.2)$$

where $c_1, c_2, \ldots c_n \vee \mathbf{R}$. Thus $c_1, c_2, \ldots c_n$ are real constants and are called the cost coefficient of the variables x. An equality constraint is linear when it has the form

$$a_1 x_1 + \cdots + a_n x_n = b \qquad (4.3)$$

where $a_1, a_2, \ldots a_n, b \vee \mathbf{R}$. A linear inequality constraint must have one of the following forms:

$$a_1 x_1 + \cdots + a_n x_n < b \qquad (4.4)$$

$$a_1 x_1 + \cdots + a_n x_n > b \qquad (4.5)$$

$$a_1 x_1 + \cdots + a_n x_n \leq b \qquad (4.6)$$

$$a_1 x_1 + \cdots + a_n x_n \geq b \qquad (4.7)$$

where $a_1, a_2, \ldots a_n, b \vee \mathbf{R}$.

When $f(x_1, x_2, \ldots x_n)$ is linear and W is determined by a system of linear equations and inequalities, the mathematical programming problem is a linear programming problem.

4.5.2.1 Linear Programming

Linear programming is a special case of mathematical programming used to achieve the best outcome in a mathematical model whose requirements are represented by linear relationships. It is an applicable technique for the optimization of a linear objective function, subject to linear equality and linear inequality constraints. Its feasible region is a set defined as the intersection of finitely many half spaces, each of which is defined by a linear inequality. Its objective function is a real-valued linear function defined on this polyhedron. A linear programming algorithm finds a point in the polyhedron where this function has the smallest (or largest) value if such a point exists (Dantzig, 1963).

Linear programs are problems that can be expressed in the following form:

$$\text{maximize } \mathbf{c}^\mathrm{T}\mathbf{x} \tag{4.8}$$

subject to

$$\mathbf{Ax} \le \mathbf{b} \text{ and } \mathbf{x} \ge 0$$

where $\mathbf{x}$ represents the vector of unknown variables, $\mathbf{c}$ and $\mathbf{b}$ are vectors of known coefficients, A is a known matrix of coefficients, and $^\mathrm{T}$ is the matrix transpose. The expression $\mathbf{c}^\mathrm{T}\mathbf{x}$ is called the objective function. The inequalities $\mathbf{Ax} \le \mathbf{b}$ and $\mathbf{x} \ge 0$ are the constraints, which specify a convex polytope over which the objective function is to be optimized.

Linear programming problems can be converted into an augmented form to apply the common form of the simplex algorithm. This form introduces nonnegative slack variables to replace inequalities with equalities in the constraints. The problems can then be written in the following block matrix form:

Maximize z:

$$\begin{bmatrix} 1 & -\mathbf{c}^\mathrm{T} & 0 \\ 0 & \mathbf{A} & \mathbf{I} \end{bmatrix} \begin{bmatrix} z \\ \mathbf{x} \\ \mathbf{s} \end{bmatrix} = \begin{bmatrix} 0 \\ \mathbf{b} \end{bmatrix} \quad \mathbf{x} \ge 0, \mathbf{s} \ge 0 \tag{4.9}$$

where $\mathbf{s}$ are the newly introduced slack variables, and z is the variable to be maximized.

4.5.3 Lagrange Multipliers

Lagrange multipliers can be used for optimization problems with equality constraints in an open form such as:

$$\text{maximize } f(x, y) \tag{4.10}$$

subject to

$$g(x, y) = 0. \tag{4.11}$$

To solve these problems explicitly, we need both f and g to have continuous first partial derivatives. A new variable (λ) called a

Lagrange multiplier is introduced and we solve the Lagrange function defined by:

$$\mathscr{L}(x, y, \lambda) = f(x, y) - \lambda g(x, y) \qquad (4.12)$$

If $f(x_0, y_0)$ is a maximum of $f(x, y)$ for the original constrained problem, then there exists λ_0 such that (x_0, y_0, λ_0) is a stationary point for the Lagrange function identified by the condition where all partial derivatives of $\mathscr{L}$ equal zero. However, not all stationary points yield a solution of the original problem. Thus the method of Lagrange multipliers yields a necessary condition for optimality in constrained problems (Lasdon, 1970). Sufficient conditions for a minimum or maximum also exist.

To solve a Lagrange function find:

$$\nabla_{x,y,\lambda} \mathscr{L}(x, y, \lambda) = 0 \qquad (4.13)$$

or

$$\nabla_{x,y} f(x, y) = -\lambda \nabla_{x,y} g(x, y) \qquad (4.14)$$

This defines a system of three equations with three unknowns.

The Lagrange method generalizes to functions with n variables.

$$\nabla_{x_1,\ldots,x_n,\lambda} \mathscr{L}(x_1, \ldots, x_n, \lambda) = 0 \qquad (4.15)$$

which requires solving $n + 1$ equations for $n + 1$ unknowns.

The constrained extrema of f are critical points of the Lagrangian $\mathscr{L}$, but they are not necessarily local extrema of $\mathscr{L}$.

The Lagrangian may be reformulated as a Hamiltonian, in which case the solutions are local minima for the Hamiltonian. This is done in optimal control theory, in the form of Pontryagin's minimum principle.

The fact that solutions of the Lagrangian are not necessarily extrema poses difficulties for numerical optimization. This can be addressed by computing the magnitude of the gradient, as the zeros of the magnitude are necessarily local minima.

4.5.4 Nonlinear

The iterative methods used to solve problems of nonlinear programming (NLP) may be classified according to whether they evaluate Hessians, gradients, or only function values. Evaluating Hessians and gradients improves the rate of convergence, but with computational complexity of each iteration. In some cases, the computational complexity may be excessively high.

One criterion for optimizers is the number of required function evaluations, as this is often a large computational effort and usually much more effort than within the optimizer itself, which operates over the N variables to be solved. The derivatives provide detailed information for such optimizers, but are even harder to calculate since approximating the gradient takes at least $N + 1$ function evaluations. For approximations of the second derivatives collected in the Hessian matrix, the number of function evaluations is in the order of N^2. Newton's method requires the second-order derivatives, so for each iteration the number of function calls is in the order of N^2, but for a simpler pure gradient optimizer it is only N. Gradient optimizers usually need more iterations than Newton's algorithm. The best method with respect to the number of function calls depends on the problem itself.

A synopsis of the various methods follows.

4.5.4.1 Hessian Methods

The two methods that evaluate Hessians or approximate Hessians using finite differences are: Newton's method (Deuflhard, 2004) and SQP.

Newton's method to find zeroes of a function of g multiple variables is given by:

$$x_{n+1} = x_n - \left[\mathbf{J}_g(x_n)\right]^{-1} g(x_n) \qquad (4.16)$$

where $[\mathbf{J}_g(x_n)]^{-1}$ is the left inverse of the Jacobian matrix $\mathbf{J}_g(x_n)$ of g evaluated for x_n.

SQP is a Newton-based method developed for small- to medium-scale constrained problems. Some versions can handle large-dimensional problems.

SQP methods apply when the objective function and the constraints are twice continuously differentiable. The method solves a sequence of optimization subproblems, each of which optimizes a quadratic model of the objective subject to a linearization of the constraints. If the problem is unconstrained, then the method reduces to Newton's method for finding a point where the gradient of the objective vanishes. If the problem has only equality constraints, then the method is equivalent to applying Newton's method to the first-order optimality conditions, or Karush—Kuhn—Tucker (KKT) conditions (Karush, 1939; Kuhn and Tucker, 1951), of the problem.

The KKT conditions (also known as the Kuhn—Tucker conditions) are first-order necessary conditions for a solution in NLP to be optimal, provided that some regularity conditions are satisfied. Allowing inequality constraints, the KKT approach to NLP generalizes the method of Lagrange multipliers, which allows only equality constraints. The system of equations corresponding to the KKT conditions is usually not solved directly, except in the few special cases where a closed-form solution can be derived analytically. In general, many optimization algorithms can be interpreted as methods for numerically solving the KKT system of equations (Boyd and Vandenberghe, 2004).

4.5.4.2 Gradient Methods

Methods that evaluate gradients or approximate gradients using finite differences or subgradients include:

- Quasi–Newton methods
- Conjugate gradient methods
- Interior point methods
- Gradient descent
- Subgradient methods
- Simultaneous perturbation stochastic approximation

Quasi–Newton methods are often employed for iterative methods for medium-large problems. Any method that replaces the exact Jacobian $Jg(x_n)$ with an approximation is a quasi–Newton method. The chord method where $Jg(x_n)$ is replaced by $Jg(x_o)$ for all iterations, for instance, is an example.

Quasi–Newton methods are used to either find zeroes or local maxima and minima of functions, as an alternative to Newton's method. They can be used if the Jacobian or Hessian is unavailable or is too expensive to compute at every iteration. Newton's method requires the Jacobian to search for zeros, or the Hessian for finding extrema.

Broyden's method (Broyden, 1965) is commonly used to find extrema that can also be applied to find zeroes. Other methods that can be used are the Column Updating Method (Martinéz, 1993), the Inverse Column Updating Method (Martinez and Zambaldi, 1992), the Quasi–Newton Least Squares Method (Haelterman et al., 2009), and the Quasi–Newton Inverse Least Squares Method (Degroote et al., 2008).

Conjugate gradient methods (Hestene and Stiefel, 1952) are iterative methods for large problems.

Interior point methods are a large classification of methods for constrained optimization. Some interior point methods use only gradient and subgradient information, whereas others require the evaluation of Hessians. The class of primal-dual path-following interior point methods (Potra and Wright, 2000) is considered the most successful. Mehrotra's predictor-corrector algorithm (Mehrotra, 1992) provides the basis for most implementations of this class of methods.

Gradient descent or steepest descent method is only used for finding approximate solutions of enormous problems.

Subgradient is an interative method for nondifferentiable functions using generalized gradients (Shor, 1985). These methods are similar to conjugate—gradient methods.

The simultaneous perturbation stochastic approximation (SPSA) method for stochastic optimization uses random gradient approximation. SPSA is a descent method capable of finding global minima. Its main feature is the gradient approximation that requires only two measurements of the objective function, regardless of the dimension of the optimization problem (Bhatnagar et al., 2013).

4.5.4.3 Evaluation of Function Values

If a problem is continuously differentiable, then gradients can be approximated using finite differences, in which case a gradient-based method can be used; otherwise interpolation or pattern search methods are used.

Interpolation is the approximation of a complicated function by a simple function. A few known data points from the original function can be used to create an interpolation based on a simpler function. When a simple function is used to estimate data points from the original, interpolation errors are usually present; however, depending on the problem domain and the interpolation method used, the gain in simplicity may be of greater value than the resultant loss in precision.

Pattern search optimization methods are also known as direct-search, derivative-free, or black-box methods. Hooke and Jeeves (1961) first developed the method. A convergent pattern-search method was proposed by Yu (1979), who proved that it converged using the theory of positive bases. The golden section search is a special case of pattern search for finding the extremum of a strictly unimodal function by successively narrowing the range of values inside which the extremum is known to exist.

4.5.4.4 Quadratic Programming

Quadratic programming optimizes a quadratic function of several variables subject to linear constraints on these variables. The quadratic programming problem with n variables and m constraints can be formulated as follows (Nocedal and Wright, 2006).

Given:
- a real-valued, n-dimensional vector $\mathbf{c}$,
- an $n \times n$ dimensional real symmetric matrix Q,
- an $n \times n$ dimensional real matrix A, and
- an m dimensional real vector $\mathbf{b}$,

the objective of quadratic programming is to find an n dimensional vector $\mathbf{x}$ that minimizes

$$\frac{1}{2}\mathbf{x}^{\mathrm{T}}Q\mathbf{x} + \mathbf{c}^{\mathrm{T}}\mathbf{x} \qquad (4.17)$$

subject to

$$Ax \leq b$$

where x^T denotes the vector transpose of x. The notation $Ax \leq b$ means that every entry of the vector Ax is less than or equal to the corresponding entry of the vector b.

A related programming problem, quadratically constrained quadratic programming, can be posed by adding quadratic constraints on the variables.

4.5.5 Discrete

An integer programming problem is characterized by a condition where some or all of the variables are restricted to be integers. If the objective function and the constraints (other than the integer constraints) are linear, then this is classified as mixed-integer linear programming (MILP).

Many real world problems involve both discrete decisions and nonlinear system dynamics. These decision problems lead to mixed-integer nonlinear programming (MINLP) problems that combine the combinatorial difficulty of optimizing over discrete variable sets with the challenges of handling nonlinear functions (Nemhauser and Wolsey, 1988).

4.5.5.1 Mixed Integer Linear Programming

MILP involves problems in which only some of the variables are constrained to be integers, whereas other variables are allowed to be nonintegers.

Mixed Integer Programming problems are of the form:

$$\text{minimize } c^T x \tag{4.18}$$

subject to

$Ax = b$ (linear constraints)

$l \leq x \leq u$ (bound constraints)

some of all x must take integer values (integrality constraints)

The integrality constraints allow MILP models to capture the discrete nature of some decisions. For example, a variable whose values are restricted to 0 or 1, called a binary variable, can be used to decide whether or not some action is taken.

An MILP solver can also be used for models with a quadratic objective and/or quadratic constraints:

$$\text{minimize } x^T Q x + q^T x \tag{4.19}$$

subject to

$A\mathbf{x} = \mathbf{b}$ (linear constraints)

$l \leq \mathbf{x} \leq u$ (bound constraints)

$\mathbf{x}^T Q_i \mathbf{x} + q_i^T \mathbf{x} \leq b_i$ (quadratic constraints)

some of all x_j must take integer values (integrality constraints).

Mixed Integer Program (MIP) models with a quadratic objective but without quadratic constraints are called Mixed Integer Quadratic Programming problems. MIP models with quadratic constraints are called Mixed Integer Quadratically Constrained Programming problems. Models without any quadratic features are often referred to as MILP problems.

4.5.5.2 Mixed Integer Nonlinear Programming

Many optimal decision problems involve both discrete decisions and nonlinear system dynamics that affect the quality of the final design or plan. MINLP problems combine the combinatorial difficulty of optimizing over discrete variable sets with the challenges of handling nonlinear functions. MINLP is one of the most general modeling paradigms in optimization and includes both NLP and MILP as subproblems (Belotti et al., 2012).

MINLPs are expressed as:

$$\text{minimize } f(x) \tag{4.20}$$

subject to

$$c(x) \leq 0, x \vee X, x_i \vee Z \quad \text{and} \quad \forall_i \vee I$$

where f and c are twice continuously differentiable functions, $X \subset R^n$ is a bounded polyhedral set, and $I \subseteq \{1,\ldots,n\}$ is the index set of integer variables. More general constraints, such as equality constraints, or lower and upper bounds $l \leq c(x) \leq u$, can be included. More general discrete constraints that are not integers can be modeled by using so-called special-ordered sets of type I (Beale and Tomlin, 1970; Beale and Forrest, 1976).

Eq. (4.20) is a nondeterministic polynomial-time hard combinatorial problem, because it includes MILP (Kannan and Monma, 1978), and its solution typically requires searching enormous search trees. Worse, non-convex integer optimization problems are in general undecidable (Jeroslow, 1973). Eq. (4.20) is decidable when X is compact or by assuming that the problem functions are convex.

Most solution methods for MINLP apply some form of tree search of which there are two broad classes of methods: single-tree and multitree methods. Classical single-tree methods include nonlinear branch-and-bound (Lawler and Woods, 1966) and branch-and-cut methods (Tawarmalani and Sahinidis, 2005), whereas classical multitree methods include outer approximation and Benders decomposition. The most efficient class of methods for convex MINLP are hybrid methods that combine the strengths of both classes of classical techniques. Two such strategies are the two-phase method and the diving method (Eckstein, 1994; Linderoth and Savelsbergh, 1999; Achterberg, 2005).

Nonconvex MINLPs pose additional challenges, because they contain nonconvex functions in the objective or the constraints; hence, even when the integer variables are relaxed to be continuous, the feasible region is generally nonconvex, resulting in many local minima. A range of approaches have been developed to tackle this challenging class of problems, including piecewise linear approximations, generic strategies for obtaining convex re-laxations nonconvex functions, spatial branch-and-bound methods, and other techniques that exploit particular types of nonconvex structures to obtain improved convex relaxations.

Heuristic techniques obtain a good feasible solution in situations where the search tree has grown too large or we require real-time solutions. Another approach is mixed-integer optimal control that adds systems of ordinary differential equations to MINLP (Sager, 2005).

4.6 ADVANCED OPTIMIZATION TECHNIQUES

Stochastic optimization methods generate and use random variables. For stochastic problems, the random variables appear in the formulation of the optimization problem itself, which involve random objective functions or random constraints. Some stochastic optimization methods use random iterates for the solution (Spall, 2003). Stochastic optimization methods generalize methods for deterministic problems.

Partly random input data arise in such areas as real-time estimation and control, simulation-based optimization where Monte Carlo simulations estimate actual systems (Fu, 2002; Campi, 2008) and problems where there is random error in the measurements of the criterion. In such cases, knowledge that the function values are contaminated by randomness (noise) leads naturally to algorithms that use statistical inference tools to estimate the "true" values of the function and/or make statistically optimal decisions about the next steps. Examples of stochastic methods include:

- stochastic approximation (Robbins and Monro, 1951)
- stochastic gradient descent
- finite difference (Kiefer and Wolfowitz, 1952)
- simultaneous perturbation (Spall, 1992)
- scenario optimization

Metaheuristics are probably the most complex advanced optimization techniques. Metaheuristics sample a set of solutions, which is too large to be completely sampled and may make few assumptions about the optimization problem being solved, and so they may be usable for a variety of problems (Blum, 2003).

Compared with optimization algorithms and iterative methods, metaheuristics do not guarantee that a globally optimal solution can be found on some class of problems. Many metaheuristics implement some form of stochastic optimization, so that the solution found depends on the set of random variables generated (Bianchi et al., 2009). By searching over a large set of feasible solutions, metaheuristics can often find good solutions with less computational effort than optimization algorithms, iterative methods, or simple heuristics. As such, they are useful approaches for optimization problems.

Metaheuristic algorithms of engineering optimization include simulated annealing, genetic algorithms, particle swarm optimization, ant colony algorithm, bee algorithm, harmony search, firefly algorithm, and many others.

4.6.1 Simulated Annealing

There are certain optimization problems that become unmanageable using combinatorial methods as the number of objects becomes large. For these problems, there is a very effective practical algorithm called simulated annealing (thus named because it mimics the process undergone by misplaced atoms in a metal when it is heated and then slowly cooled). Although this technique is unlikely to find the optimum solution, it can often find a very good solution, even in the presence of noisy data.

Simulated annealing is a probabilistic technique for approximating the global optimum of a given function. Specifically, it is a metaheuristic to approximate global optimization in a large search space. It is often used when the search space is discrete. For problems where finding the precise global optimum is less important than finding an acceptable local optimum in a fixed amount of time, simulated annealing may be preferable to alternatives such as brute-force search or gradient descent.

Simulated annealing interprets slow cooling as a slow decrease in the probability of accepting worse solutions as it explores the solution space. Accepting worse solutions is a fundamental property of metaheuristics because it allows for a more extensive search for the optimal solution. The method is an adaptation of the Metropolis—Hastings algorithm, a Monte Carlo method to generate sample states of a thermodynamic system, invented by M.N. Rosenbluth (Metropolis et al., 1953).

Such worse solutions or "bad" trades are allowed using the criterion that:

$$e^{-\Delta D/T} > R(0, 1) \tag{4.21}$$

where ΔD is the change of distance implied by the trade (negative for a "good" trade; positive for a "bad" trade), T is a "synthetic temperature," and $R(0, 1)$ is a random number in the interval $[0,1]$. D is a cost function and corresponds to the free energy in the case of annealing a metal (in which case the temperature parameter would actually be the kT, where k is Boltzmann's constant and T is the physical temperature, in the Kelvin absolute temperature scale). If T is large, many "bad" trades are accepted, and a large part of solution space is accessed. Objects to be traded are generally chosen randomly, although techniques that are more sophisticated can be used.

The second step, again by analogy with annealing of a metal, is to lower the "temperature." After making many trades and observing that the cost function declines only slowly, one lowers the temperature, and thus limits the size of allowed "bad" trades. After lowering the temperature several times to a low value, one may "quench" the process by accepting only "good" trades to find the local minimum of the cost function. There are various "annealing schedules" for lowering the temperature, but the results are generally not very sensitive.

There is another faster strategy called threshold acceptance (Dueck and Scheuer, 1990). In this strategy, all good trades are accepted, as are any bad trades that raise the cost function by less than a fixed threshold. The threshold is then periodically lowered, just as the temperature is lowered in annealing. This eliminates exponentiation and random number generation in the Boltzmann criterion. As a result, this approach can be faster in computer simulations.

4.6.2 Genetic Algorithms

A genetic algorithm is a method for solving both constrained and unconstrained optimization problems based on a natural selection process that mimics biological evolution. The algorithm repeatedly modifies a population of individual solutions.

4.7 DYNAMIC OPTIMIZATION

Dynamic optimization is the process of finding the optimal control profile of one or more control variables or control parameters of a system. Optimality is defined as the minimization or maximization of an objective function without violating the specified constraints.

Optimal control deals with the problem of finding a control law for a given system such that a certain optimality criterion results. A control problem includes a cost function containing state and control variables. Optimal control is a set of differential equations describing the paths of the control variables that minimize the cost functional. The optimal control can be derived using Pontryagin's maximum principle (Ross, 2009) or by solving the Hamilton–Jacobi–Bellman equation (a sufficient condition). Model predictive control uses optimal control formulas in its structure.

The objective is to minimize the continuous time cost function expressed as:

$$J = \Phi[\mathbf{x}(t_0), t_0, \mathbf{x}(t_f), t_f] + \int_{t_0}^{t_f} \mathcal{L}[\mathbf{x}(t), \mathbf{u}(t), t]dt \qquad (4.22)$$

subject to the first-order dynamic constraints (the state equation)

$$\dot{x}(t) = a[\mathbf{x}(t), \mathbf{u}(t), t], \qquad (4.23)$$

the algebraic path constraints

$$b[\mathbf{x}(t), \mathbf{u}(t), t] \leq 0 \qquad (4.24)$$

and the boundary conditions

$$\Phi[\mathbf{x}(t_0), t_0, \mathbf{x}(t_f), t_f] \qquad (4.25)$$

where $\mathbf{x}(t)$ is the state, $\mathbf{u}(t)$ is the control, t is the independent variable (usually time), t_0 is the initial time, and t_f is the terminal time. The terms Φ and $\mathcal{L}$ are the end-point cost and Lagrangian, respectively. The path constraints are in general inequality constraints and thus may not be zero at the optimal solution.

The optimal control problem may have multiple solutions. Thus it is most often the case that any solution $\mathbf{x}^*(t^*), \mathbf{u}(t^*), t^*$ to the optimal control problem a local minimum.

4.7.1 Linear Quadratic Control

A special case of the general nonlinear optimal control problem given in the previous section is the linear quadric (LQ) optimal control problem. The LQ problem is stated as follows. Minimize the quadratic continuous-time cost function expressed as:

$$J = \frac{1}{2} \mathbf{x}^T(t_f)S_f\mathbf{x}(t_f) + \int_{t_0}^{t_f} [\mathbf{x}^T(t)\mathbf{Q}(t)\mathbf{x}(t) + \mathbf{u}^T(t)\mathbf{R}(t)\mathbf{u}(t)]dt \qquad (4.26)$$

Subject to the linear first-order dynamic constraints

$$\dot{\mathbf{x}}(t) = \mathbf{A}(t)\mathbf{x}(t) + \mathbf{B}(t)\mathbf{u}(t) \tag{4.27}$$

and the initial condition

$$\mathbf{x}(t_0) = \mathbf{x}_0 \tag{4.28}$$

A particular form of the LQ problem that arises in many control system problems is the linear quadratic regulator (LQR) where all of the matrices, **A, B, Q** and **R**, are constant, the initial time is arbitrarily set to zero, and the terminal time is taken in the limit $t_f \rightarrow \infty$, which is known as infinite horizon. The LQR problem is formulated as follows.

Minimize the infinite horizon quadratic continuous-time cost function expressed as:

$$J = \frac{1}{2} \int_0^\infty [\mathbf{x}^T(t)\mathbf{Q}(t)\mathbf{x}(t) + \mathbf{u}^T(t)\mathbf{R}(t)\mathbf{u}(t)] \, dt \tag{4.29}$$

subject to the linear time-invariant first-order dynamic constraints

$$\dot{\mathbf{x}}(t) = \mathbf{A}\mathbf{x}(t) + \mathbf{B}\mathbf{u}(t) \tag{4.30}$$

and the initial condition (Eq. 4.28).

In the finite-horizon case the matrices are restricted in that **Q** and **R** are positive semidefinite and positive definite, respectively. In the infinite-horizon case, the matrices **Q** and **R** are not only positive-semidefinite and positive-definite, respectively, but also constant. These additional restrictions on **Q** and **R** in the infinite-horizon case are enforced to ensure that the cost function remains positive. Furthermore, to ensure that the cost function is bounded, the additional restriction is imposed that the pair (**A,B**) is controllable. The LQ or LQR cost function attempts to minimize the control energy.

In the infinite horizon problem the operator is driving the system to zero state and hence driving the output of the system to zero. The problem of driving the output to a desired nonzero level can be solved after the zero output solution is found. The LQ or LQR optimal control has the feedback form:

$$\mathbf{u}(t) = -\mathbf{K}(t)\mathbf{x}(t) \tag{4.31}$$

where $\mathbf{K}(t)$ is a properly dimensioned matrix, given as

$$\mathbf{K}(t) = \mathbf{R}^{-1}\mathbf{B}^T\mathbf{S}(t) \tag{4.32}$$

and $\mathbf{S}(t)$ is the solution of the differential Riccati equation. The differential Riccati equation is given as

$$\dot{\mathbf{S}}(t) = -\mathbf{S}(t)\mathbf{A} - \mathbf{A}^T\mathbf{S}(t) - \mathbf{A}^T\mathbf{S}(t) + \mathbf{S}(t)\mathbf{B}\mathbf{R}^{-1}\mathbf{B}^T\mathbf{S}(t) - \mathbf{Q} \quad (4.33)$$

For the finite horizon LQ problem, the Riccati equation is integrated backward in time using the terminal boundary condition:

$$\mathbf{S}(t_f) = \mathbf{S}_f \quad (4.34)$$

For the infinite horizon LQR problem, the differential Riccati equation is replaced with the algebraic Riccati equation (ARE) given as:

$$0 = -\mathbf{S}\mathbf{A} - \mathbf{A}^T\mathbf{S} - \mathbf{A}^T\mathbf{S} + \mathbf{S}\mathbf{B}\mathbf{R}^{-1}\mathbf{B}^T\mathbf{S} - \mathbf{Q} \quad (4.35)$$

Understanding that the ARE arises from infinite horizon problem, the matrices $\mathbf{A}$, $\mathbf{B}$, $\mathbf{Q}$, and $\mathbf{R}$ are all constant. There are multiple solutions to the ARE and the positive definite or positive semidefinite solution is the one that is used to compute the feedback gain (Kalman, 1960).

4.7.2 Numerical Methods for Optimal Control

Optimal control problems are generally nonlinear and, therefore, generally unlike the linear-quadratic optimal control problem do not have analytic solutions. It is necessary to employ numerical methods to solve optimal control problems. In the early years of optimal control the favored approach for solving optimal control problems was that of indirect methods employing the calculus of variations to obtain the first-order optimality conditions. These conditions result in a multipoint boundary value problem. This boundary value problem has a special structure because it arises from taking the derivative of a Hamiltonian. Thus the resulting dynamical system is a Hamiltonian system of the form:

$$\dot{\mathbf{x}} = \partial H/\partial \lambda \quad (4.36)$$

$$\dot{\lambda} = -\partial H/\partial \mathbf{x} \quad (4.37)$$

where

$$H = \mathscr{L} + \lambda^T \mathbf{a} - \mu^T \mathbf{b} \quad (4.38)$$

is the augmented Hamiltonian and in an indirect method, the boundary value problem is solved using the appropriate boundary or transversality conditions. In an indirect method the state and adjoint (λ) are determined and the resulting solution is readily verified to be an extremal trajectory. The disadvantage of indirect methods is that the boundary value problem is often extremely difficult to solve for problems that span large time intervals or problems with interior point constraints.

Direct methods were developed in the 1980s where the state and/or control are approximated using an appropriate function such as polynomial approximation or piecewise constant parameterization. Then, the co-efficients of the function approximations are treated as optimization variables and the problem is transcribed to a nonlinear optimization problem of the form:

$$\text{Minimize } F(z) \tag{4.39}$$

subject to the constraints

$$\mathbf{g(z) = 0}$$
$$\mathbf{h(z) \leq 0}$$

Depending on the type of direct method employed, the size of the nonlinear optimization problem may be quite large. The reason for the relative ease of computation, particularly of a direct collocation method, is that the NLP is sparse and many well-known software programs exist to solve large sparse NLPs. As a result, the range of problems that can be solved via direct methods (particularly direct collocation methods) is significantly larger than the range of problems that can be solved via in-direct methods.

4.8 REAL-TIME OPTIMIZATION

Building the underlying model for optimization requires a true rigorous, first principles modeling tool usually taking an equation-oriented approach, where the underlying relationships are described by equations representing the process physics and chemistry. The resulting set of equations is solved transparent to the user.

RTO refers to the online economic optimization of a process plant, or a section of a process plant. An opportunity for implementing RTO exists when the following criteria are met:

- Adjustable optimization variables exist after higher priority safety, product quality, and production rate objectives have been achieved.
- The profit changes significantly as values of the optimization variables are changed.
- Disturbances occur frequently enough for real-time adjustments to be required.
- Determining the proper values for the optimization variables is too complex to be achieved by selecting from several standard operating procedures.

Real-time, online adaptive control of processing systems is possible when the control algorithms include the ability to build multidimensional response surfaces that represent the process being controlled. These response surfaces, or knowledge capes, change constantly as processing conditions, process inputs, and system parameters change, providing a real-time basis for process control and optimization.

RTO applications have continued to develop in their formative years with 100 or so worldwide large-scale processing applications. RTO systems are frequently layered on top of an advanced process control (APC) system, producing economic benefits using highly detailed thermodynamic, kinetic, and process models, and nonlinear optimization. An APC system typically pushes material and energy balances to increase feed and preferred products with some elements of linear optimization, whereas RTO systems can trade yield, recovery, and efficiency among disparate pieces of equipment.

Processors frequently use RTO applications for off-line studies because they provide a valuable resource for debottlenecking and evaluating changes in feed, catalyst, equipment configuration, operating modes, and chemical costs. Processing RTO has been hampered by the lack of reactor models for major processing units, property estimation techniques for hydrocarbon streams, and the availability of equipment models. Continued technology advancement has removed many of these hurdles, however, and the number of reported RTO successes continues to grow.

RTO systems perform the following main functions:
- Steady-state detection—Monitor the plant's operation and determine if the plant is sufficiently steady for optimization using steady-state models.
- Reconciliation/parameter estimation—Collect operating conditions and reconcile the plant-wide model determining the value of the parameters that represent the current state of the plant.
- Optimization—Collect the present operating limits (constraints) imposed and solve the optimization problem to find the set of operating conditions that result in the most profitable operation.
- Update set points—Implement the results by downloading the optimized set points to the historian for use by the control system.

A good RTO system utilizes the best process engineering technology and operates on a continuous basis. The system constantly solves the appropriate optimization problem for the plant in its present state of performance and as presently constrained.

A typical system consists of an efficient (fast) equation solver/optimizer "engine," coupled with robust, detailed, mechanistic (not correlation based) equipment models, and a complete graphical interface that contains a real-time scheduling (RTS) system and an external data interface to the process computer. Three primary components of a fully integrated graphical interface are shown in Figs. 4.1–4.3.

The RTO model is composed of separate models for each major piece of equipment. These separate models are integrated and are solved simultaneously. The simultaneous solution (rather than sequential) approach allows for solution of large-scale, highly integrated problems that would be difficult or impossible to solve using sequential techniques offered by many flow sheet vendors.

An RTO system will determine the plant optimum operation in terms of a set of optimized set points. These will then be implemented via the control system.

4.8.1 Physical Properties

All of the process models for a rigorous RTO system use mixture physical properties, such as enthalpy, K-values, compressibility, vapor pressure, and entropy.

Equations of state, such as Soave–Redlich–Kwong or Peng–Robinson, are used for fugacities, enthalpy, entropy, and compressibility of the hydrocarbon streams. The enthalpy datum is based on methods such as the enthalpy of formation from the elements at absolute zero temperature. This allows the enthalpy routines to calculate heats of reaction as well as sensible heat changes. Steam and water properties are calculated using routines based on standards such as the National Institute of Standards and Technology.

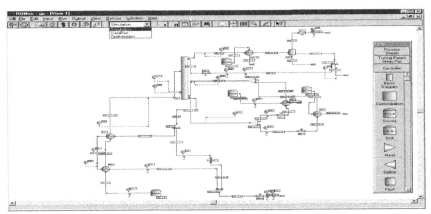

Figure 4.1 Real-time optimization interactive model.

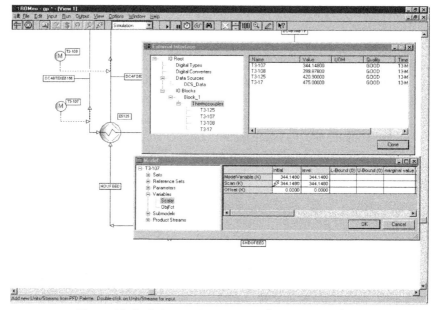

Figure 4.2 Real-time optimization configuration interface.

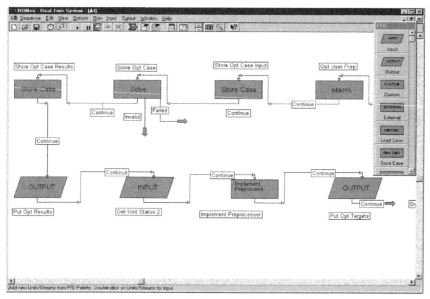

Figure 4.3 Real-time optimization real-time scheduling interface.

4.8.2 Optimization Models

Models used in the optimization system must be robust, easily specified (in an automated manner) to represent changing plant situations and must solve efficiently. These models must be able to fit observed operating conditions, have sufficient fidelity to predict the interactions among independent and dependent variables and represent operating limits or constraints. State of the art models meet these requirements by using residual format $(0 = f(x))$, open equations, fundamental, mechanistic relationships, and by incorporating meaningful parameters within the models. These state-of-the-art systems provide the highly efficient equation solver/optimizer, and the interface functionality that automates sensing the plant's operating conditions and situations, and automate posing and solving the appropriate parameter and optimization cases.

Most plant models are standard models. Sometimes custom models are created specifically for the equipment of a unit. All the models use thermodynamic property routines for enthalpy, vapor—liquid equilibrium, and entropy information.

Rotating equipment models such as compressor, pump, engine, gas turbine, and steam turbine models contain, along with all the thermodynamic relationships, the expected performance relationships for the specific equipment modeled.

4.8.3 Optimization Objective Function

The objective function maximized by a high-level optimization system is the net plant profit. This is calculated as product values minus feed costs minus utility costs, i.e., the P − F − U form. When appropriately constrained, this objective function solves either the "maximize profit" or "minimize operating cost" optimization problem. Economic values are required for each of the products, feeds, and utilities. The value of each stream is derived from the composition-weighted sum of its components.

Economic values for feeds, products, and utilities in the optimization system are reset on a regular basis for best performance.

4.8.4 Custom Models

Often custom models must be incorporated into a standard optimization package to predict proprietary processes and solvents. The custom model may be incorporated with special properties packages or integrated into the system in a semiopen approach where the iteration criterion is handled by the optimization system but the actual kinetics or thermodynamic equations remain the same as the off-line custom model.

Proprietary gas sweetening solvent formulations may be the most common example of a custom model in the natural gas industry.

4.8.5 Fractionators

Fractionators are modeled using tray-to-tray distillation. Heating and cooling effects, as well as product qualities, are considered. Column pressure is typically a key optimization parameter. Temperature measurements are used to determine heat transfer coefficients for condensers and reboilers.

4.8.6 Absorbers and Strippers

These units will be modeled using tray-to-tray distillation, if required. Component splitters may be used to simplify the flowsheet where acceptable when these units have little effect on the optimum.

4.8.7 Compression Model

The main optimization set point variables for the compressors are typically suction or discharge pressures. Important capacity constraints are maximum and minimum speed for the driver, maximum current for electric motors, and maximum power for steam turbines, gas turbines, and engines. Sometimes maximum torque will be considered for engines.

A multistage compressor model consists of models for a series of compressor stages, interstage coolers, and adiabatic flashes. Drivers are included for each compressor machine.

For each single compressor stage, the inlet charge gas conditions (pressure, temperature, flow rate, and composition) and the discharge pressure specification are used with the manufacturer's compressor performance curves to predict the outlet temperature and compressor speed. The power required for the compression is calculated from the inlet and outlet conditions.

The first step in the development of the compression model is to fit the manufacturer's performance curves for polytropic head and efficiency to polynomials in suction volume and speed. The equations for each compressor stage (or wheel, if wheel information is available) take the form:

$$E_p = A \times N^2 + B \times N + C \times N \times V_s + D \times V_s + E \times V_s^2 + F \quad (4.40)$$

$$W_p = a \times N^2 + b \times N + c \times N \times V_s + d \times V_s + e \times V_s^2 + f \quad (4.41)$$

where V_s = suction volume flow rate; N = compressor speed; Ep = stage or wheel polytropic efficiency; W_p = stage or wheel polytropic head, and A-F and a-f are correlation constants.

The polytropic head change across the stage or wheel can also be calculated from the integral of VdP from suction pressure to discharge pressure. This integration can be performed by substituting $V = Z \times R \times T/P$ and integrating by finite difference approximation:

$$W_p = R \times \ln(P_d/P_s) \times ((Z_s \times T_s) + (Z_d \times T_d))/2 \qquad (4.42)$$

where R = gas constant; P_d = discharge pressure; P_s = suction pressure; Z_s = compressibility at suction conditions; T_s = absolute suction temperature; Z_d = compressibility at discharge conditions; T_d = absolute discharge temperature.

For simplicity, the above-mentioned equations are formulated in terms of a single integration step between suction and discharge, but in the actual implementation, each stage or wheel is divided into at least five sections to describe the true profile. The enthalpy change across the stage or wheel can be calculated from the inlet and outlet conditions, the polytropic head change, and the polytropic efficiency.

This analysis results in three simultaneous equations and three unknowns for each integration step. The unknowns are the discharge temperature, discharge pressure, and enthalpy change for each step except the last. The known discharge pressure for the last step is related to the speed of the machine. In state-of-the-art approaches, all of the integration steps are solved simultaneously.

Measured discharge pressures, speeds, and temperatures are used in the parameter estimation run to update the intercept terms in the polynomials used to represent the manufacturer's polytropic head and efficiency curves. These parameters represent the differences between the actual performance and expected/design performance. As compressors foul, these parameters show increasing deviation from expected performance. It is this difference that has significant meaning since an absolute calculation of efficiency at any moment in time can vary with feed rate and several other factors, which dilute the meaning of the value. By showing a difference from design, we get a true measure of the equipment performance and how it degrades over time.

The discharge flow from each stage is sent through a heat exchanger model coupled with an adiabatic flash. The heat transfer coefficient for each exchanger is based on the measured suction temperature to the next stage, corrected for addition of any recycle streams. Suction and/or discharge flows are measured to fit heat loss terms in the interstage flash drums.

4.8.8 Distillation Calculations

Standard tray-to-tray distillation models are used for distillation calculations in an optimization system. The "actual" number of trays is used wherever possible and performance is adjusted via efficiency. This allows the model to more accurately represent the plant in a way that is understandable to a plant operator.

All distillation models predict column-loading constraints accurately as targets are changed. Condenser and reboiler duties are also calculated for predicting utility requirements and exchanger limitations.

4.8.8.1 Tray-to-Tray Distillation Method

An equation-based tray-to-tray distillation method is based on mass, heat, and vapor—liquid equilibrium balances on each physical tray. Fig. 4.4 is a typical distillation column tray.

Component mole balances for each component i on tray j are:

$$F_i Z_{ij} + L_{i+1} X_{(i+1)j} + V_{(i-1)} Y_{(i-1)j} - (L_i + LP_i) X_{ij} - (V_i + VP_i) Y_{ij} = 0$$
(4.43)

The overall mole balance is:

$$F_i + L_{i+1} + V_{i-1} - (L_i + LP_i) - (V_i - VP_i) = 0 \qquad (4.44)$$

The vapor—liquid equilibrium definition is written in terms of the liquid mole fractions and K-values:

$$Y_{ij} = K_{ij}(X_{ij}, T_{il}, P_i) \times X_{ij}, \qquad (4.45)$$

where T_{il} is the liquid temperature on tray i.

The requirement that the mole fractions balance is expressed as:

$$\sum [X_{ij}] - \sum [K_{ij}(X_{ij}, T_{il}, P_i) \times X_{ij}] = 0 \qquad (4.46)$$

The Murphree vapor tray efficiency, which accounts for differences between the equilibrium vapor composition [$K_{ij}(X_{ij}, T_{il}, P_i) \times X_{ij}$] and the actual mixed vapor composition [Y_{ij}] leaving the tray, is an important adjustable parameter. It is defined as:

$$E_{ij} = (Y_{ij} - Y_{ij-1})/(K_{ij}(X_{ij}, T_{il}, P_i) \times X_{ij} - Y_{ij-1}) \qquad (4.47)$$

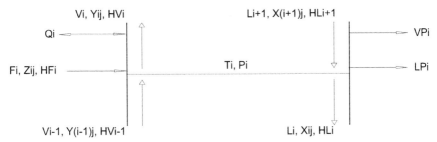

Figure 4.4 Mass transfer operations in a tray distillation column.

The vapor and liquid enthalpies are defined as:

$$HL_i - HL(X_{ij}, T_{il}, P_i) = 0, \text{ and} \qquad (4.48)$$

$$HV_i - HV(Y_{ij}, T_{iv}, P_i) = 0, \qquad (4.49)$$

where T_{iv} is the vapor temperature on tray i.

The overall tray heat balance is:

$$F_iHF_i + L_{i+1}HL_{i+1} + V_{i-1}HV_{i-1} - (L_i + LP_i)HL_i \\ - (V_i + VP_i)HV_i + Q_i = 0 \qquad (4.50)$$

The liquid and vapor flow rates on the capacity limiting trays are used in conjunction with tray loading calculations to predict column pressure drops. The vapor area factor in the loading correlation is parameterized to enable matching calculated column differential pressure against the measured differential pressure. Upper limits are placed on this calculated differential pressure during the optimization case to ensure avoidance of flooding.

Condenser and reboiler limitations are handled by placing constraints on the minimum approach temperature in these exchangers or on the maximum heat transfer area in the case of variable level exchangers. Heat transfer coefficients are calculated for the exchangers in the parameter case.

4.8.8.2 Demethanizer

Set point variables for a cryogenic demethanizer are typically the bottoms methane or carbon dioxide content and the overhead pressure. Constraint variables include overhead condensing capability and minimum column temperatures. The expansion impacts on the demethanizer and the overhead ethane losses are key profit variables.

The demethanizer is modeled with a tray-to-tray distillation model. Rigorous K-values for the demethanizer column are recommended due to nonlinear vapor–liquid equilibrium relationships. Overall heat transfer coefficients for the reboilers and feed chillers are calculated from temperature and flow data.

4.8.8.3 Deethanizer

The set point variables in the deethanizer are the overhead pressure, the overhead propane content, and the bottoms ethane content. Any cooling medium (such as refrigeration, cooling water, or air) and heat medium (such as steam or hot oil) requirements of the deethanizer and the propane loss in the overhead are the major profit considerations.

The deethanizer is modeled with a tray-to-tray distillation model. Parameters include column pressure, column pressure drop, propane content in the deethanizer overhead, reflux flows, deethanizer bottoms ethane content, and bottoms draw rate. Overall heat transfer coefficients are calculated for the exchangers. Any preheaters are included in the deethanizer scope.

4.8.8.4 Depropanizer

The set point variables for the depropanizer are the overhead butane content and column pressure. Key constraint variables are the propane in the debutanizer overhead, the overhead exchanger capacities, and column pressure drop (flooding). Any cooling and heat medium usage requirements of the depropanizer are the major profit considerations.

The depropanizer is modeled with tray-to-tray distillation models. Parameters considered include column feed temperature, column pressure, column pressure drop, butane content in the tops, reflux flow, bottoms propane content and bottoms draw rate. Overall heat transfer coefficients are calculated for the reboiler and condenser. A column capacity factor is also parameterized.

4.8.8.5 Debutanizer

The set point variables for the debutanizer are the overhead pentanes and heavier content as well as column pressure. Key constraint variables are the propane in the debutanizer overhead, the overhead exchanger capacities, and column pressure drop (flooding). Any heating and cooling medium usage requirements of the debutanizer are the major profit considerations.

The debutanizer is modeled with a tray-to-tray distillation model. Parameters considered include column feed temperature, column pressure, column pressure drop, pentanes and heavier content in the overhead, reflux flow, bottoms butanes content, and bottoms draw rate. Overall heat transfer coefficients are calculated for the reboiler and condenser. A column capacity factor is also parameterized.

4.8.8.6 Butanes Splitter

The set point variables for the butanes splitter is the column pressure, normal butane in overhead, and isobutane in the bottoms. The constraint variables of interest are reboiler and condenser loading and product specifications. The profit variables of interest are the isobutane losses in the bottoms normal butane stream.

The butanes splitter is modeled with a tray-to-tray distillation model. The parameters considered include column pressure, column differential pressure, normal butane in the isobutane product, bottoms isobutane concentration, reflux flow rate, bottoms flow rate, product draws, and bottom reboiler flow rate. Heat transfer coefficients for the exchangers are calculated.

4.8.9 Refrigeration Models

The main set point variable for refrigeration machines is the first-stage suction pressure.

Refrigeration system models relate refrigeration heat loads to compressor power. The compressor portions of the refrigeration models use the same basic equations as compressor models discussed earlier. Refrigeration systems should use the appropriate composition and will have a component mixture for any makeup gas.

The measured compressor suction flows and the heat exchange duties calculated by the individual unit models are used to determine the total refrigeration loads. The refrigerant vapor flows generated by these loads are calculated based on the enthalpy difference between each refrigerant level. Exchanger models of the refrigerant condensers are used to predict compressor discharge pressures.

4.8.10 Demethanizer Feed Chilling Models

The demethanizer feed chilling system is modeled as a network of heat exchangers and flash drums. These models are used to predict the flow rates and compositions of the demethanizer feeds. The effects of changing demethanizer system pressures and flow rates are predicted.

The demethanizer feed drum temperatures and feed flow rates are measured to fit the fractions of heat removed from the feed gas by each exchanger section. Flow and temperature measurements on the cold stream

side allow the fraction of feed gas heat rejected to each stream to be estimated. The two sides of the feed exchangers are coupled through an overall
heat balance. The inlet and outlet temperatures from the refrigerant and
other process exchangers are used to fit overall heat transfer factors in the
parameter case. A pressure drop model for the feed gas path is also included.

4.8.11 Steam and Cooling Water System Models

Heat and material balance models of the steam system are developed. These
models include detailed representations of the boilers.

The cooling water system will be modeled with heat exchangers,
mixers, and splitters to allow for constraining the cooling water temperature and its effects on the operation of distillation columns and
compressors.

4.8.12 Turbines

A turbine model, as shown in Fig. 4.5, is used for steam turbines (back
pressure, condensing, or extraction/condensing) or any expander in which
the performance relationship can be expressed using the following
equation:

$$\text{Design Power} = A + B \times \text{Mass Flow} + C \times (\text{Mass Flow})^2 \\ + D \times (\text{Mass Flow})^3 \tag{4.51}$$

Backpressure and condensing turbine expected performances are usually
presented as essentially linear relationships between power and steam flow.
Extracting/condensing turbine expected performance relationships are
typically presented as power versus throttle steam flow, at various extraction
steam flows. This kind of performance "map" can be separated into two
relationships of the form given earlier, one representing the extraction
section and the other the condensing section. An extraction/condensing
turbine can be thought of as two turbines in series, with part of the
extraction section flow going to the condensing section.

Design power refers to expected power from performance "maps" that
are at specific design inlet pressure, inlet temperature, and exhaust pressure.
Expected power is the power expected from a turbine operating at other
than design conditions. The design power is adjusted by the "power factor," as illustrated:

$$\text{Expected Power} = \text{Design Power} \times \text{Power Factor} \tag{4.52}$$

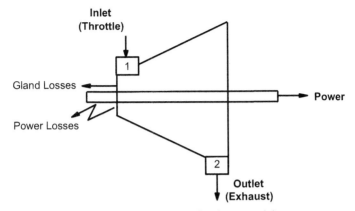

Figure 4.5 Schematic of turbine model.

$$\text{Power Factor} = \frac{\Delta \text{ Isentropic Enthalpy @ actual conditions}}{\Delta \text{ Isentropic Enthalpy @ design conditions}} \qquad (4.53)$$

$$\text{Brake (Shaf) Power} = \text{Expected Power} + \text{Power Bias} \qquad (4.54)$$

Design power and expected power are equal if actual expanding fluid (i.e., steam) conditions are the same as design conditions. The power bias can be parameterized using the exhaust temperature for backpressure turbines, or the extraction section of an extraction/condensing turbine. In both these cases the exhaust steam is superheated (single phase). For condensing turbines or the condensing section of an extraction/condensing turbine, the measured steam flow is calculated since the temperature of the two-phase exhaust cannot be used. The fraction vapor of the two-phase exhaust is determined by energy balance. The total energy demand can be determined from the compressor (or other driven power consumer). The power extracted from the extraction section can be determined from the throttle steam flow and the inlet and outlet (single phase) conditions. The power extracted from the condensing section is just the difference between the total demand and the extraction section power.

The power loss is taken from the steam and affects its outlet conditions, but is not transferred to the shaft, or brake power.

To calculate the change of enthalpies needed for the power factor and for the energy balance, fluid conditions are calculated for each turbine section at the inlet (throttle), outlet (exhaust), and at inlet entropy and outlet pressure for both actual and design conditions. The nomenclature associated with the exhaust fluid conditions required for the enthalpy change calculations is:

Exh IDM = Exhaust **I**sentropic at **D**esign inlet conditions for vapor/liquid **M**ixture.

Exh IDV = Exhaust **I**sentropic at **D**esign inlet conditions for **V**apor.

Exh IDL = Exhaust **I**sentropic at **D**esign inlet conditions for **L**iquid.

Exh IAM = Exhaust **I**sentropic at **A**ctual inlet conditions for vapor/liquid **M**ixture.

Exh IAV = Exhaust **I**sentropic at **A**ctual inlet conditions for **V**apor.

Exh IAL = Exhaust **I**sentropic at **A**ctual inlet conditions for **L**iquid.

For condensing turbines or the condensing section of extraction/condensing turbines the exhaust pressure is used to specify the model. The heat transfer coefficient is updated by this measurement. The exhaust pressure is free to move in the optimization cases, since the actual pressure moves as the condenser calculates the pressure required to condense the steam sent to the condensing turbine or condensing section.

The maximum mass flow at reference (usually design) conditions is used to predict the maximum mass flow at actual conditions as an additional constraint on the turbine performance. Sonic flow relationships are used for this prediction. This is the maximum flow through the inlet nozzles when the inlet steam chest valves are wide open. The nozzle area is not needed if the maximum flow at a set of reference conditions is known. Vendors usually list the maximum flow, and not the nozzle area.

The model also includes the gland steam flow, which is used to counterbalance the axial thrust on the turbine shaft, and which is lost through labyrinth seals. The gland steam does not contribute to the shaft power.

Typically, turbine performance is not considered to be a function of speed. Consequently speed is a variable that has no effect on the solution. However, turbine performance is typically a very weak function of speed over a wide range, and its effect on performance is not typically presented on the expected performance "maps." The possibility of adding the weak effect of speed on performance will be considered when the model is being built.

4.8.13 Plant Model Integration

After the plant section models have been developed, they are integrated into the overall plant model. All of the interconnecting streams are specified and checked and a consistent set of variable specifications developed. At this stage, the overall validity of the plant model is checked using off-line plant data.

Reconciliation/parameter and optimization cases are run and the results are checked for accuracy and reasonableness. A material balance model is included in the plant model integration work to confirm an absolute model material balance closure. This material balance will include a furnace area balance, a recovery area balance, and an overall balance. In addition, the plant integration allows the objective function to be tested and validated with connections to all feed, product, and utility variables.

It is important to have the engineers that will be responsible for commissioning the optimization system involved in the project during plant model integration. A thorough understanding of the plant model is imperative for a smooth implementation of the online system.

4.8.14 Model Fidelity and Measurement Errors

In an optimization system as described here, neither the models nor the measurements need to be absolutely perfect for the system to work well and to deliver significant improvement in profitability. An online optimization system, as shown in Fig. 4.6, continuously receives feedback from real-time measurements. The model parameters are updated before each optimization so that the models fit the plant and the optimum set points calculated are valid and can be confidently implemented. Without this constant feedback of plant measurements and regular updating of the plant model, the optimization solution might not be feasible. The fidelity of the models and the accuracy of the measurements are reflected in trends of the parameters. During commissioning of the optimization system, the best available measurements are identified by analyzing many parameter cases, running with real-time data, before closing the loop.

If, for example, significant heat balance discrepancies exist between process side and fuel/flue gas side measurements in a furnace, that discrepancy can be handled by determining the bias required on the fuel gas flow measurement(s) to satisfy the heat balance. That bias could be a parameter that is updated before each optimization. The variation of the bias over time (its trend) would be monitored. If this parameter varies significantly, the furnace model and other measurements used by it would be investigated thoroughly. Alternate measurements used to "drive" the parameter case solution would be investigated as well. The outcome of this analysis is that the best available measurements are selected, and the model

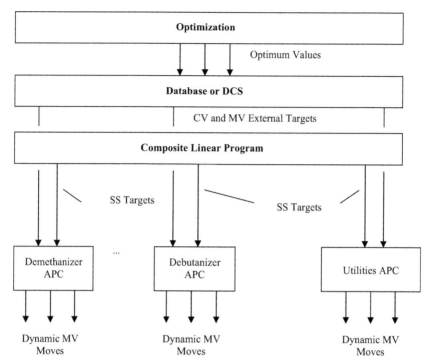

Figure 4.6 Advanced process control (APC) with optimization solution architecture. *MV*, Manipulated variable; *SS*, steady state.

relationships are thoroughly investigated to ensure that all significant relationships are included. This analysis is a standard and required step in building the optimization system.

Validity checking is an integral part and is built into an online optimization system. It is used to screen out gross errors. If alternate measurements are available, the validity checkers can use these when primary measurements are unavailable or bad. Generic validity checkering takes care of common errors, whereas custom validity checking can respond to site-specific situations. Measurements can be designated as critical, so if they are unavailable, the optimization cycle will be directed to monitor for steady state, and will only complete its cycle when the measurement becomes available. Validity checking has several features designed to keep the online service factor of the optimization system high in the face of imperfect measurements. Another feature of the online system is that the measurements used to drive the solution of the parameter case are averages over a specified time window (usually 1 h) so that measurement noise is suppressed.

The better the measurements and the better the models, the better an RTO can fully exploit the process equipment and consequently the more potential profit is realized.

Processing RTO in the future will likely include wider applications driven by demonstrated benefits, reduced implementation costs, and acceptance as a best practice. RTO applications are also becoming tightly intertwined with planning and scheduling systems, in which they can update models and calculate stream costs.

Evolving technologies changing the value proposition for refining RTO include:

- Detailed kinetic models for all major processing units and configurations, proven by reported applications.
- Optimization technology improvements that incorporate robust solvers, integer variables, and the capability to handle increasing problem sizes. Today's technology can handle applications with several hundred thousand equations; a typical refining application has 100 measurements, more than 100,000 variables, and 25 outputs.
- Greater integration with higher-level systems including shared models and reconciled measurement data.
- Multiunit optimization that leverages shared resources between process units and continues to lead toward rigorous refinery-wide optimization.
- Computing technology improvements, which have already shifted RTO from minicomputers to personal computers, and that will allow more solutions/day and more complex formulations. Solver and computing improvements will eventually lead to true dynamic optimization.
- Application and model building tools, operating graphical user interfaces, and sustained performance technologies that will lower cost, improve benefits, and remove other hurdles.

4.9 PROCESS OPTIMIZATION CASE STUDY

One outstanding opportunity for process optimization was the determination of the correct levels of ethane and propane recovery recognizing that ethane margins were often negative while propane margins were very positive (Felizzola et al., 2015).

The plant of interest installed a cryogenic natural gas liquids recovery unit designed with an ethane rejection mode capable of 75% ethane rejection with approximately 95% propane recovery. Operating data revealed average performance of about 88% propane recovery and 50% ethane recovery.

Various methods and technologies were considered to assist operators with controlling the plant at more profitable conditions. It was determined that an accurate plant economic model that could calculate the optimum amount of propane to recover and ethane to reject subject to current conditions of feed rate, feed composition, equipment capabilities, and contractual constraints was the best way to capture this opportunity. This system would be capable of determining optimal key controller set points in the current environment of ethane rejection, and also any other economic conditions.

The solution selected not only provided outstanding value for determining the optimal trade-off between ethane rejection and propane recovery for all conditions encountered by the plant but also was able to immediately recognize when ethane (and other components are profitable to recover). The solution also understands the impact that each key variable such as feed gas splits, expander speed, plant pressure profile, and demethanizer temperatures have on the relative recovery of each component in the feed gas to the cryogenic unit.

RTO technology was able to capture the opportunity of assisting the operators as they cope with ethane rejection requirements by:

- collecting real-time plant data as well as current specific pricing and nominations from the commercial departments,
- running a reconciled first principles model of the plant process,
- determining an optimum profit (in this case dollars per thousand cubic feet of feed gas subject to equipment and contractual constraints such as residue gas delivery nominations and heating value, and
- reporting the optimum key variable values corresponding to controller set points.

In many cases, APC is used to transfer the optimum calculated variables as remote set points to the plant control system. However, in this case it was decided to provide the suggested set points as advice for the plant operator to act upon.

The optimization technology was selected as a system to provide operator decision support for today's ethane rejection challenges as well as operating the plant in the most profitable fashion in the future under any economic scenario.

Another example on a gas processing complex in Norway has been reported (Kovach et al., 2010).

REFERENCES

Achterberg, T., 2005. SCIP — a Framework to Integrate Constraint and Mixed Integer Programming. Technical Report ZIB-Report 04-19. Konrad-Zuse-Zentrum für Informationstechnik, Berlin, Takustr, 7, Berlin.

Beale, E., Tomlin, J., 1970. Special facilities in a general mathematical programming system for non-convex problems using ordered sets of variables. In: Lawrence, J. (Ed.), Proceedings of the 5th International Conference on Operations Research, Pages 447—454, Venice, Italy.

Beale, E.M.L., Forrest, J.J.H., 1976. Global optimization using special ordered sets. Mathematical Programming 10, 52—69.

Belotti, P., Kirches, C., Leyffer, S., Linderoth, J., Luedtke, J., Mahajan, A., 2012. Mixed-Integer Nonlinear Optimization, Argonne National Laboratory, Mathematics and Computer Science Division.

Bhatnagar, S., Prasad, H.L., Prashanth, L.A., 2013. Stochastic Recursive Algorithms for Optimization: Simultaneous Perturbation Methods. Springe.

Bianchi, L., Dorigo, M., Gambardella, L., Gutjahr, W., 2009. A survey on metaheuristics for stochastic combinatorial optimization. Natural Computing: an International Journal 8 (2), 239—287.

Blum, C., Roli, A., 2003. Metaheuristics in combinatorial optimization: overview and conceptual comparison. ACM Computing Surveys 35 (3), 268—308.

Boyd, S., Vandenberghe, L., 2004. Convex Optimization. Cambridge University Press, Cambridge.

Broyden, C.G., 1965. A class of methods for solving nonlinear simultaneous equations. Mathematics of Computation (American Mathematical Society) 19 (92).

Campi, M.C., Garatti, S., 2008. The exact feasibility of randomized solutions of uncertain convex programs. SIAM Journal on Optimization 19 (3), 1211—1230.

Dantzig, G.B., 1963. Linear Programming and Extensions, Princeton. Princeton University Press, NJ.

Degroote, J., Haelterman, R., Annerel, S., Swillens, A., Segers, P., Vierendeels, J., 2008. An interface quasi-Newton algorithm for partitioned simulation of fluid-structure interaction. In: Hartmann, S., Meister, A., Schfer, M., Turek, S. (Eds.), Proceedings of the International Workshop on Fluid-Structure Interaction. Theory, Numerics and Applications. Kassel University Press, Germany.

Deuflhard, P., 2004. Newton methods for nonlinear problems. Affine invariance and adaptive algorithms. In: Springer Series in Computational Mathematics, vol. 35. Springer, Berlin.

Dueck, G., Scheuer, T., 1990. Threshold accepting: a general purpose optimization algorithm appearing superior to simulated annealing. Journal of Computational Physics 90, 161—175.

Eckstein, J., 1994. Parallel branch-and-bound algorithms for general mixed integer programming on the CM-5. SIAM Journal on Optimization 4, 794—814.

Felizzola, C., Morrison, R., Poe, W., 2015. Coping with rejection. In: 95th GPA Annual Convention.

Fletcher, R., 1987. Practical Methods of Optimization, second ed. John Wiley & Sons, New York.

Fu, M.C., 2002. Optimization for simulation: theory vs. practice. Informs Journal on Computing 14 (3), 192—227.

Haelterman, R., Degroote, J., Van Heule, D., Vierendeels, J., 2009. The quasi-Newton least squares method: a new and fast secant method analyzed for linear systems. SIAM Journal on Numerical Analysis 47 (3), 2347—2368.

Hestenes, M., Stiefel, E., 1952. Methods of conjugate gradients for solving linear systems. Journal of Research of the National Bureau of Standards 49 (6).

Hooke, R., Jeeves, T.A., 1961. 'Direct search' solution of numerical and statistical problems. Journal of the Association for Computing Machinery (ACM) 8 (2), 212–229.

Jeroslow, R.G., 1973. There cannot be any algorithm for integer programming with quadratic constraints. Operations Research 21 (1), 221–224.

Jeter, M., 1986. Mathematical Programming – an Introduction to Optimization. Marcel Dekker, Inc.

Kalman, R., 1960. A new approach to linear filtering and prediction problems. Journal of Basic Engineering 82, 34–45. Transactions of the ASME.

Kannan, R., Monma, C., 1978. On the computational complexity of integer programming problems. In: Henn, R., Korte, B., Oettli, W. (Eds.), Optimization and Operations Research, Volume 157 of Lecture Notes in Economics and Mathematical Systems, Pages 161–172. Springer.

Karush, W., 1939. Minima of Functions of Several Variables with Inequalities as Side Constraints (M.Sc. Dissertation). Dept. of Mathematics, Univ. of Chicago, Chicago, Illinois.

Kiefer, J., Wolfowitz, J., 1952. Stochastic estimation of the maximum of a regression function. Annals of Mathematical Statistics 23 (3), 462–466.

Kovach, J.W., Meyer, K., Nielson, S.-A., Pedersen, B., Thaule, S.B., March 21–24, 2010. The role of online models in planning and optimization of gas processing facilities: challenges and benefits. In: 89th GPA Annual Convention, Austin, TX, USA.

Kuhn, H.W., Tucker, A.W., 1951. Nonlinear programming. In: Proceedings of 2nd Berkeley Symposium. Berkeley: University of California Press.

Lasdon, L., 1970. Optimization Theory for Large Systems. Macmillan Series in Operations Research. The Macmillan Company, New York.

Lawler, E.L., Woods, D.E., 1966. Branch-and-bound methods: a survey. Operations Research 14 (4), 699–719.

Linderoth, J.T., Savelsbergh, M.W.P., 1999. A computational study of search strategies in mixed integer programming. Informs Journal on Computing 11, 173–187.

Martinéz, J.M., 1993. On the convergence of the column-updating method. Communications on Pure and Applied Mathematics 12, 83–94.

Martinez, J.M., Zambaldi, M.C., 1992. An inverse column-updating method for solving large scale nonlinear systems of equations. Optimization Methods and Software 1 (2), 129–140.

Mehrotra, S., 1992. On the implementation of a primal–dual interior point method. SIAM Journal on Optimization 2 (4), 575–601.

Metropolis, N., Rosenbluth, A., Rosenbluth, M., Teller, A., Teller, E., 1953. Equation of state calculations by fast computing machines. The Journal of Chemical Physics 21 (6), 1087.

Nelder, J., Mead, R., 1965. A simplex method for function minimization. Computer Journal 7, 308–313.

Nemhauser, G., Wolsey, L.A., 1988. Integer and Combinatorial Optimization. John Wiley and Sons, New York.

Nocedal, J., Wright, S., 2006. Numerical Optimization, second ed. Springer-Verlag, Berlin, New York.

Potra, F., Wright, S.J., 2000. Interior-point methods. Journal of Computational and Applied Mathematics 124 (1–2), 281–302.

Robbins, H., Monro, S., 1951. A stochastic approximation method. Annals of Mathematical Statistics 22 (3), 400–407.

Ross, I.M., 2009. A Primer on Pontryagin's Principle in Optimal Control. Collegiate Publishers.

Sager, S., 2005. Numerical Methods for Mixed-integer Optimal Control Problems. Der andere Verlag, Tönning, Lübeck, Marburg, ISBN 3-89959-416-9.

Shor, N., 1985. Minimization Methods for Non-differentiable Functions. Springer-Verlag.

Sorensen, D., 1982. Newton's Method with a Model Trust Region Modification. SIAM Journal on Numerical Analysis 19 (2), 409–426.

Spall, J.C., 1992. Multivariate stochastic approximation using a simultaneous perturbation gradient approximation. IEEE Transactions on Automatic Control 37 (3), 332–341.

Spall, J.C., 2003. Introduction to Stochastic Search and Optimization. Wiley.

Tawarmalani, M., Sahinidis, N.V., 2005. A polyhedral branch-and-cut approach to global optimization. Mathematical Programming 103 (2), 225–249.

Yang, X., 2010. Engineering Optimization – an Introduction with Metaheuristic Applications. John Wiley & Sons, Inc.

Yu, W., 1979. Positive basis and a class of direct search techniques. Scientia Sinica [Zhongguo Kexue] 53–68.

APPENDIX A

Basic Principles of Control Valves

Control valves regulate the rate of fluid flow as the force of an actuator changes the position of a valve plug or disk. Some considerations for the design of a control valve include:

- no external leakage under fluid design pressure,
- adequate capacity at fluid design flow,
- minimal erosion and corrosion,
- withstand design temperature, and
- incorporate appropriate end connections and actuator attachment means.

The control valve assembly consists of the valve body, actuator, positioner, air sets, transducers, limit switches, and relays.

A.1 TYPES OF CONTROL VALVES

The most popular types of control valves are globe valves and rotary valves. The different configurations of these types of control valves are given in the following sections.

A.1.1 Globe Valves

A linear closure member, one or more ports, and a globular-shaped cavity around the port characterize globe valves. Globe valves can be further classified as:

- Single ported
- Double ported
- Angle style
- Three way
- Balanced cage guided
- High-capacity cage guided

Most globe valves have a selection of valves, cages, or plugs that can be interchanged to modify the inherent flow characteristic (Emerson Process Management, 2005).

A.1.1.1 Single Port

Single port is the most common valve body style due to its simplicity and is available in globe, angle, bar stock, forged, and split construction. Single port valves are generally specified for stringent shutoff requirements and handle most services. Generally, single-port valves are less than 4 inches in size, but are sometimes used up to 8 inches in size with high-thrust actuators (Fig. A.1).

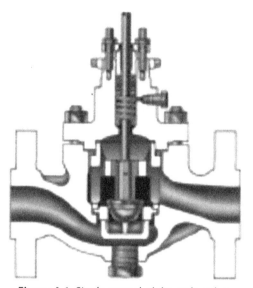

Figure A.1 Single-ported globe-style valve.

Port-guided valve bodies are usually limited to 150 psi maximum pressure drop and are susceptible to velocity-induced vibration, but with special construction consideration these valves are often used in natural gas service. They are typically provided with screwed in seat rings, which might be difficult to remove after use (Fig. A.2; Emerson Process Management, 2005).

A.1.1.2 Double Ported

With double-ported dynamic forces on the plug are balanced because flow opens one port as it closes the other. This balanced force permits a smaller actuator than required for a single-ported valve of similar capacity. These valves are usually 4 inches or larger and have higher capacity than single-ported valves of the same size (Fig A.3).

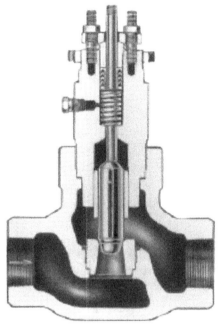

Figure A.2 High-pressure single-ported globe-style valve.

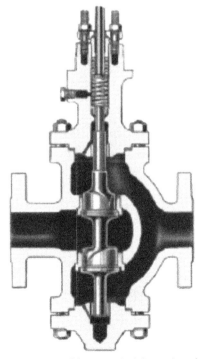

Figure A.3 Double-ported globe-style valve.

Double-ported valves can be provided in either push down to open or close configurations. These valve bodies are typically used for on—off or low-pressure throttling service. They are good for viscous fluids with dirt and contaminants, but only meet class II or III leak classifications (Emerson Process Management, 2005).

A.1.1.3 Angle Style

Angle valves are usually single ported. Common services are boiler feed-water, heater drain service, and in piping schemes where space is a pre-mium. These valves can serve as an elbow. Cage style, screwed–in seat rings, expanded outlet connections, restricted trim, and outlet liners for reduction of erosion damage are common constructions (Fig. A.4; Emerson Process Management, 2005).

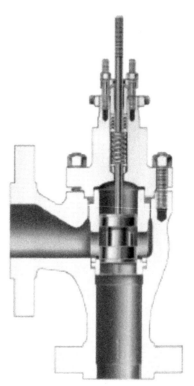

Figure A.4 Angle-style valve.

A.1.1.4 Three Way

Three-way valves provide three pipeline connections for flow mixing and flow splitting service. Most designs use cage-style trim for positive valve plug guiding and ease of maintenance. These valves will install with all standard end connections and can include high temperature trim materials (Fig. A.5; Emerson Process Management, 2005).

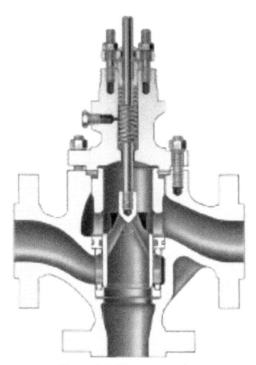

Figure A.5 Three-way valve.

A.1.1.5 Balanced Cage Guided

This valve style uses one set ring and provides the advantages of a balanced valve plug associated with a double-port valve. The cage-style trim provides valve plug guidance and seat ring retention. A sliding piston ring-type seal prevents leakage of the higher pressure upstream fluid into the downstream lower pressure system by allowing the downstream pressure to act on the top and bottom sides of the valve plug. Reduction of the unbalanced static force permits smaller actuators than those necessary for conventional single-ported valves. A multitude of trim choices allows several flow characteristics, noise attenuation, and anticavitation. Balanced cage-guided valves are available in sizes up to 20 inches and pressure rating to Class 2500 (Fig. A.6; Emerson Process Management, 2005).

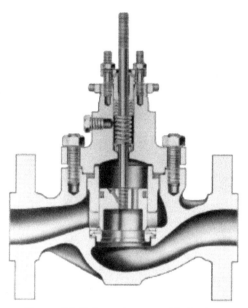

Figure A.6 Balanced cage-guided valve.

A.1.1.6 High-Capacity Cage Guided

High-capacity cage-guided valves are designed for noise applications such as high-pressure gas reducing stations where sonic velocities may be encountered. This design incorporates oversize end connections with a streamlined flow path. Noise abatement trim reduces overall noise levels as much as 35 dB. High-capacity cage-guided valves are available in sizes up to 20 inches and pressure rating to Class 600 (Fig. A.7; Emerson Process Management, 2005).

A.1.2 Rotary Valves

Rotary shaft control valves contain a flow closure member that rotates in the flowstream to control the capacity of the valve. Rotary valve styles include butterfly, v-notch ball, eccentric disk, and eccentric plug (Emerson Process Management, 2005).

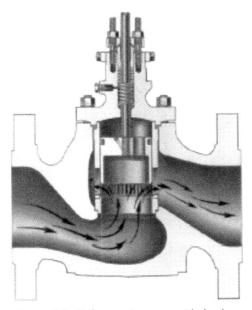

Figure A.7 High-capacity cage-guided valve.

A.1.2.1 Butterfly Valves

The characteristics of butterfly valves are (Fig A.8; Emerson Process Management, 2005):

- the valve body requires minimal installation space
- high capacity with low pressure loss
- conventional disks provide throttling for up to 60 degrees disk rotation with dynamically streamlined disks allowing up to 90 degrees disk rotation
- bodies mate with standard raised face flanges
- operating torques are high for large valves or high pressure drop
- good shutoff and corrosion protection with standard liners
- sizes up to 72 inches
- approximately equal percentage flow characteristic
- useful for throttling or on—off control

A.1.2.2 V-Notch Ball Valves

V-notch ball valves have good rangeability (greater than 300:1), control, and shutoff capability with an equal percentage flow characteristic. Other characteristics of V-notch ball vales include (Fig. A.9; Emerson Process Management, 2005):

- low pressure drop due to straight-through flow design
- suitability for erosive or viscous fluids

W4841

Figure A.8 High-performance butterfly control valve.

W8172-2

Figure A.9 V-notch ball control valve.

- compatibility with standard diaphragm or piston rotary actuators
- flanged or flangeless end connections

A.1.2.3 Eccentric Disk Valves

Characteristics of eccentric disk valves include (Fig. A.10; Emerson Process Management, 2005):
- effective throttling control
- linear flow characteristic through 90 degrees of disk rotation
- minimal seal wear
- sizes up to 24 inches
- compatibility with standard diaphragm or piston rotary actuators
- reverse flow reduces capacity
- imprecise throttling control
- approximately one-third the control range of a ball or globe-style valve

Figure A.10 Eccentric disk control valve.

A.1.2.4 Eccentric Plug Valves

Characteristics of eccentric plug valves include (Fig. A.11; Emerson Process Management, 2005):
- erosion resistance
- prolonged seat life

Figure A.11 Sectional of eccentric plug control valve.

- good throttling performance
- tight shutoff in forward or reverse flow
- flanged or flangeless design

A.2 CONTROL VALVE PERFORMANCE

The ability of controls to reduce process variability depends on many factors. The control valve is one of the most important factors for good response to disturbances with the most important design considerations including:
- dead band,
- actuator and positioner design,
- valve response time,
- valve type and characterization, and
- valve sizing.

A.2.1 Dead Band

Dead band is the range through which an input signal can be varied without initiating an observable change in the output signal. Referring to Fig. A.12 the controller output (CO) is the input to the valve assembly and

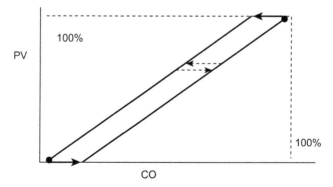

Figure A.12 Process dead band.

the process variable (PV) is the output (Emerson Process Management, 2005).

Dead band is typically expressed as a percentage of the input span.

Some of the causes of dead band are:

- friction,
- backlash (slack in a mechanical connection),
- shaft wind-up in rotary valves, and
- relay dead zone.

Seals, packing, and actuators mainly cause friction. Rotary valves are more susceptible to friction than globe valves due to their inherent design. Spring and diaphragm-type actuators tend to have less friction than piston actuators.

For best performance, total dead band should be less than 1% and ideally as low as 0.25%.

A.2.2 Actuator and Positioner Design

Actuator and positioner design must be considered concurrently. The combined design affects the static performance and the dynamic response of the control valve assembly.

A.2.3 Valve Response Time

It is important that a control valve reach a specific position quickly. Valve response time considers both the static and dynamic time of the valve assembly. The static time is essentially the dead time or the time that it takes for the valve assembly to begin to move. The dynamic time is a measure of how long the actuator takes to attain 63% of the desired move once it starts moving.

Dead band can be caused by friction in the valve body, actuator, or positioner. The dead time should be kept as small as possible with a general goal of one-third of the overall response time, but certainly less than the process time constant. The dead time should be consistent in both stroking directions for ease of loop tuning.

The positioner and actuator determine the dynamic time.

A.2.4 Valve Type and Characterization

The inherent characteristic is the relationship between the valve flow capacity and the valve opening when the pressure drop across the valve is constant. Typical valve characteristics are linear, equal percentage, and quick opening.

The inherent valve gain is defined as the ratio of the change in flow to incremental change in valve travel. The linear characteristic is a constant inherent valve gain throughout its range. Quick opening valves have the greatest inherent valve gain at the lower end of the range and equal percentage valves have the greatest inherent valve gain at the upper end of the range of valve openings.

The valve flow passage geometry dictates the inherent valve characteristic. Most globe valves have a selection of valve cages and plugs that can modify the flow characteristics, whereas ball valves, butterfly valves, and eccentric plug valves cannot be easily changed.

Rangeability is the ratio of the maximum and minimum controllable flow rates.

A.2.5 Valve Sizing

A valve must be sufficient sized for the maximum required flow and allowable pressure drop, but a valve that is too large is detrimental to good process control. Best control performance occurs when the majority of the loop gain comes from the controller.

A.3 CONTROL VALVE ACTUATORS AND POSITIONERS

- Pneumatic valve actuators
- Electric valve actuators
- Hydraulic valve actuators
- Control valve positioners

A.4 CONTROL VALVE SIZING AND SELECTION

Selection of the most appropriate control valve depends on the following factors:

- Type of fluid
- Temperature, viscosity, and specific gravity of the fluid
- Maximum and minimum flow capacity
- Maximum and minimum inlet and outlet pressure
- Normal pressure drop
- Maximum permissible noise level
- Degrees of superheat
- Percentage vapor for two-phase fluids
- Inlet and outlet pipeline size and schedule

- Body material
- End connections and valve rating
- Action desired on air failure
- Instrument air supply pressure available
- Instrument signal

The following can be used as sizing and selection guidelines (Bishop et al., 2002):

- In a pumped circuit, the pressure drop allocated to the control valve should be equal to 33% of the dynamic losses in the system at the rated flow, or 15 psi, whichever is greater.
- The pressure drop allocated to a control valve in the suction or discharge line of a centrifugal compressor should be 5% of the absolute suction pressure, or 50% of the dynamic losses of the system, whichever is greater.
- In a system where static pressure moves liquid from one pressure vessel to another, the pressure drop allocated to the valve should be 10% of the lower terminal vessel pressure, or 50% of the system's dynamic losses, whichever is greater.
- Pressure drops in valves in steam lines to turbines, reboilers, and process vessels should be 10% of the design absolute pressure of the steam system, or 5 psi, whichever is greater.
- The gain on a control valve should never be less than 0.5.
- Avoid using the lower 10% and upper 20% of the valve stroke. The valve is much easier to control in the 10–80% range.
- Generally, control valve bodies are one size less than the line size. If this causes the valve body to be significantly less than the line size, which would reduce the valve's effective sizing coefficient (C_v), then do not apply this generalization.

The valve sizing coefficient (C_v) for a liquid is the number of US gallons per minute of water at 60°F that will flow through a valve with a 1 pound per square inch pressure drop (International Society of Automation, 2007).

$$C_v = \frac{q}{F_p \times \sqrt{(\Delta P)/G}} \tag{A.1}$$

where q is the volume rate of flow in gallons per minute; F_p is the piping geometry factor; ΔP is the difference between upstream and downstream absolute static pressure in pounds per square inch; G is the liquid specific gravity.

F_p, the piping geometry factor, is used if any fittings are directly attached to the inlet and outlet connections of the control valve. These values can be found in flow coefficient tables.

For compressible fluids:

$$C_v = \frac{q}{1360 \times F_p \times P_1 \times Y\sqrt{x/(G_g \times T_1 \times Z)}} \qquad \text{(A.2)}$$

where q is the volume rate of flow in standard cubic feet per hour; P_1 is the upstream absolute static pressure in pounds per square inch absolute; Y is the expansion factor as expressed in Eq. (A.3); G_g is the gas specific gravity; T_1 is the absolute upstream temperature in degrees R; Z is the compressibility factor.

$$Y = 1 - \frac{x}{3 \times F_\gamma \times x_T} \qquad \text{(A.3)}$$

where x is the ratio of pressure drop to upstream absolute static pressure; F_γ is the ratio of specific heats factor as expressed in Eq. (A.4); x_T is the rated pressure drop ratio factor.

$$\text{Specific heat ratio factor, } F_\gamma = \gamma/1.4 \qquad \text{(A.4)}$$

The pressure differential ratio factor, X_T, is available from the valve manufacturer. This pressure differential ratio factor may require reduction for any attached fittings.

If x is greater than $F_\gamma \times X_T$, then choked flow is predicted. The differential pressure at which x equals $F_\gamma \times X_T$ is defined as the critical pressure drop (GPSA, 2012).

A.4.1 Flow Characteristic Selection

Following are some guidelines that are helpful in deciding which type of flow characteristic is best suited for a particular application (Bishop et al., 2002):

Select equal percentage characteristic when the major portion of the control system pressure drop is not through the valve and for temperature and pressure control loops.

Select linear characteristic in liquid level or flow control service where the pressure drop across the valve is expected to remain fairly constant and the major portion of the control system's pressure drop is through the valve.

Select quick opening characteristic for frequent on—off service, such as in batch or semi-continuous processes, or where an instantly large flow is required for safety or deluge systems.

A.4.2 Cryogenic Service Valves

Cryogen leakage is not only dangerous, but also very expensive, especially when one considers the cost to make a gas into a cryogen in the first place. Thermal swings are difficult on valves. The components of the valve will contract and expand at different rates because of different material composition or the amount of time exposed to the cryogen.

Heat gains from the environment are a constant battle when dealing with cryogens, hence the reason for valve insulation. During gas processing, we are dealing with the physical properties of gases, where the liquids do not want to be liquids at atmospheric pressure and, if allowed, can violently transform to gases.

In the normal processing of cryogens, there always is the buildup of pressure because of this heat gain from the environment and the subsequent vapor formation. There needs to be special consideration in designing the valve for these pressures.

Other big problems with cryogenic services include seat leakage potential. The linear and radial growth of the stem in relation to the body often can be underestimated. Selecting the correct valve can help avoid these problems.

A.4.2.1 Selecting a Valve for Cryogenic Service

A standard soft seal will become very hard and less pliable thus not providing the shut-off required from a soft seat. Special elastomers have been applied in these temperatures but require special loading to achieve a tight seal.

Packing is a concern in cryogenic applications because of the frost that may form on valves in cryogenic applications. Moisture from the atmosphere condenses where the temperature of the surface is below freezing and creates a layer of frost. The layer of frost on the stem is drawn through the packing as the valve is actuated causing tears and thus loss of seal strokes the stem. Extension bonnets will allow the packing box area of the control valve to be warmed by ambient temperatures, thus preventing frost from forming on the stem and packing box areas. The length of the extension bonnet depends on the application temperature and insulation requirements. The colder the application, the longer the extension bonnet required.

Materials of construction for cryogenic applications are generally CF8M body and bonnet material with 300 series stainless steel trim material. In flashing applications, hard facing might be required to combat erosion.

A.5 COMMON PROBLEMS OF CONTROL VALVES

A.5.1 Cavitation

Cavitation is an issue sometimes encountered with control valves in liquid service. As a fluid flows through the restricted area of the valve, the pressure reduces. If the pressure reduces to the vapor pressure, then bubbles will appear. Some of the pressure drop will be recovered downstream of the restricted area. The bubbles cannot exist at the increased downstream pressure and will collapse (International Society of Automation, 1995).

This collapse of the bubbles produces noise and vibration, which will eventually cause physical damage and performance problems.

A.5.2 Leakage

A certain amount of seat leakage is allowed based on the leakage class designation shown in Table A.1 (Emerson Process Management, 2005).

A.5.3 Control Valve Nonlinearities

A nonlinear flow characteristic should be linearized to obtain good control performance throughout the valve's operating range. Linearization is accomplished through a control block, function generator, or a lookup table placed between the controller and the valve or with a digital positioner.

Design of a linearizer involves determination of the flow characteristic curve of the valve operating in the actual process. A minimum of three readings of the flow or PV and CO under steady-state conditions at various CO levels spanning the entire operating range are required.

A.6 DIAGNOSING CONTROL VALVE PROBLEMS

In addition to control loop tuning, some mechanical issues occur with control valves (Ruel). These include:
- Dead band
- Backlash
- Hysteresis
- Resolution
- Stiction

When a command has been given to a control valve for a new position, the behavior of the valve will vary depending on its existing position, its direction, and the amplitude of the signal. When a simple position change is made, there is dead time due to the positioner.

Table A.1 Leak classifications

Leak class	Recommended seat load	Maximum leakage allowable	Test medium	Test pressure	Testing procedure required
I	No factory leak test required	Not applicable	Not applicable	Not applicable	Not required
II	20 pounds per lineal inch of port circumference	0.5% of rated capacity	Air or water at 50–125°F	45–60 psig or maximum operating differential, whichever is lower	Pressure applied to valve inlet with outlet open to atmosphere or to a low head loss measuring device and full normal operating thrust provided by actuator
III	40 pounds per lineal inch of port circumference	0.1% of rated capacity	Air or water at 50–125°F	45–60 psig or maximum operating differential, whichever is lower	Pressure applied to valve inlet with outlet open to atmosphere or to a low head loss measuring device and full normal operating thrust provided by actuator
IV	40 pounds per lineal inch of port circumference up to 4-3/8 inches 80 pounds per lineal inch of port circumference greater than 4-3/8 inches	0.01% of rated capacity	Air or water at 50–125°F	45–60 psig or maximum operating differential, whichever is lower	Pressure applied to valve inlet with outlet open to atmosphere or to a low head loss measuring device and full normal operating thrust provided by actuator

Continued

Table A.1 Leak classifications—cont'd

Leak class	Recommended seat load	Maximum leakage allowable	Test medium	Test pressure	Testing procedure required
V	Meal seat—100 pounds per lineal inch of port circumference plus 13.75 pounds per lineal inch per psi of shutoff pressure drop	0.0005 ml of water per inch of orifice diameter per psi differential	Water at 50–125°F	Maximum service pressure across valve plug	Pressure applied to valve inlet after filling entire body cavity and connected piping with water and stroking valve plug closed. Use net specified maximum actuator thrust. Allow time for leakage flow to stabilize
VI	Meal seat—300 pounds per lineal inch of port circumference	Refer to Table A.2	Air or nitrogen at 50–125°F	50 psig or maximum operating differential, whichever is lower	Pressure applied to valve inlet. Actuator should be adjusted to operating conditions specified with full normal closing thrust applied to valve plug seat. Allow time for leakage to stabilize and use suitable measuring device

Table A.2 Class VI maximum seat leakage allowable

Nominal port diameter	Maximum leakage allowable (mL/min)
1	0.15
1−1/2	0.30
2	0.45
2−1/2	0.60
3	0.90
4	1.70
5	4.00
6	6.75

Backlash is a form of dead band that results from a temporary discontinuity between the input and output of a device when the input of the device changes direction. Slack or looseness of a mechanical connection is a typical example. If a valve has some backlash, the loop will have a tendency to oscillate. Reducing the tuning parameters will only hide the problem and make the loop slower to respond.

The position of a valve with hysteresis will vary whether the signal increases or decreases. Hysteresis usually comes from backlash, but is can also be caused by nonlinearities such as seals or friction. Hysteresis provokes oscillations and reduces performance. When a change occurs, hysteresis will also add to the dead time.

Resolution is the minimum possible change in input required to produce a detectable change in the output when no reversal of the input takes place. Resolution is typically expressed as a percentage of the input span (Fig. A.13).

When resolution is perfect, then backlash is the sole source of dead band and hysteresis (Figs. A.14 and A.15).

Stiction is the resistance to the start of motion, usually measured as the difference between the driving force required to overcome static friction. Anything that comes in contact with the moving parts of a valve will create friction. It is not possible to move a valve until the actuator overcomes this friction. When stiction occurs, the dead time will be longer as friction must be overcome before the valve will move (Fig. A.16).

Friction of a moving object is less than when it is stationary. Stiction will prevent the stem from moving for small control input changes and then the stem moves when the force is sufficient. The result of stiction is that the force required to move the stem is greater than the force required to reach the desired stem position. In the presence of stiction, the valve movement is jumpy. Overshoot may occur if the valve moves too far. Overshoot will destabilize the loop particularly if the loop is fast.

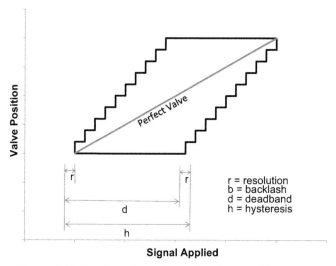

Figure A.13 Dead band, resolution, backlash, and hysteresis.

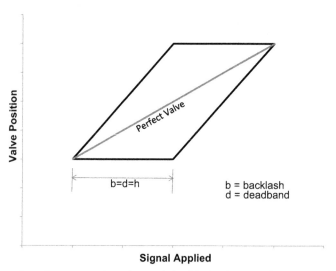

Figure A.14 Response of a valve with backlash when resolution is perfect.

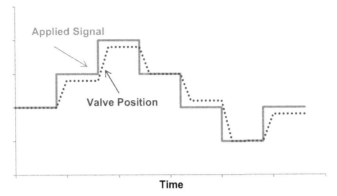

Figure A.15 Determination of the backlash in a valve. If the backlash is zero, the valve position and the process variable will go back to the previous position. Ideally, backlash in a valve is zero, but in most valves, it is near 1%. On most processes, a backlash of 2% or 3% is acceptable if the controller is not tuned too aggressively. Such backlash adds dead time to the loop and reduces performance.

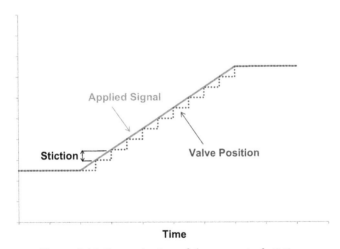

Figure A.16 Determination of the amount of stiction.

A.7 CONTROL VALVE RELIABILITY

Accurately diagnosing and correcting the control valve issues described in the previous sections obviously increases control valve reliability. Although control valve reliability is important for good process control, the most important reliability concerns are when control valves serve a safety function. In addition, studies have revealed between 2% and 16% in lost production due to control valve reliability.

A.7.1 Smart Valve Technology

First-generation smart diagnostics could only be used on out-of-service valves. These diagnostics were limited to tests such as valve signature for information on assembly friction, bench set, spring rate, and seat load; dynamic error band, which is combined hysteresis, dead band, and dynamic error; and step response to a change in input signal. All require interpretation of data by a skilled technician (Rhinehart, 2006).

With the advent of microprocessor-based valve instrumentation and sensor technology, the health of control valve assemblies became more visible. Data collected by the valve provide unprecedented diagnostics.

Smart valve technology enables fault detection and analysis while the valve is in normal operation. The diagnostics not only collect data, but also analyze the cause of a fault and suggest how best to remedy it.

Smart valve technology utilizes digital valve controllers coupled with communications protocols like HART and Foundation Fieldbus. A smart control valve provides performance metrics as well as diagnostics. These predictive diagnostics are online, in-service, and do not intrude on the process.

Digital valve controllers used with control valve assemblies can monitor valve health. Detection of a condition with the potential to affect control causes an early predictive warning, which can be used to avoid an unplanned shutdown or to schedule maintenance activity on that valve. A reported error indicates a fault is affecting the process and may require immediate action.

The diagnostics are preconfigured to collect and correlate the data, establish the cause of a problem, and recommend specific corrective action. Many different problems that impact the dynamic performance of control valves can be detected and evaluated on the most advanced digital valve controller. Some of the problems addressed include:
- instrument air leakage,
- valve assembly friction,
- process dead band or resolution,
- packing failure,
- actuator diaphragm failure,
- piston actuator seal-ring failure,
- instrument air quality issue,
- poor connections,
- filter plugging,

- incorrect bench set,
- supply pressure restrictions,
- travel deviation,
- process "buildup" on the valve trim,
- valve sticking at temperature and pressure, and
- valve assembly calibration.
  These diagnostics typically result in one of three conditions:
- No fault detected and the valve remains in service and monitoring continues.
- A warning that a fault has been detected, but control remains unaffected. This is a predictive indication that the detected problem has the potential to affect control and that future maintenance should be planned.
- An error report that a fault affecting control has been detected. These faults generally require immediate attention.

In-service diagnostics oversee some of the conditions described in the following section.

A.7.1.1 Instrument Air Leakage

Air mass flow diagnostics measure instrument air flow through the control valve assembly. Because of multiple sensors, this diagnostic can detect both positive (supply) and negative (exhaust) air mass flow from the digital valve controller. This diagnostic not only detects leaks in the actuator or related tubing, but also other problems such as leaking piston seals or damaged O-rings (Emerson Process Management, 2005).

A.7.1.2 Supply Pressure

The supply pressure diagnostic detects control valve problems related to instrument supply pressure. This diagnostic will detect both low and high supply pressure readings. In addition to checking for adequate supply pressure, this diagnostic can be used to detect and quantify droop in the air supply during large travel excursions. This is particularly helpful in identifying supply line restrictions (Emerson Process Management, 2005).

A.7.1.3 Travel Deviation and Relay Adjustment

The travel deviation diagnostic is used to monitor actuator pressure and travel deviation from set point. This diagnostic is useful in identifying a stuck control valve, active interlocks, low supply pressure, or shifts in travel calibration.

The relay adjustment diagnostic is used to monitor crossover pressure on double-acting actuators. If the crossover pressure is too low, the actuator loses stiffness, thus making the valve plug position susceptible to buffeting by fluid forces. If the crossover pressure is set too high, both chambers will be near supply, the pneumatic forces will be roughly equal, the spring force will be dominant, and the actuator will move to its spring-fail position (Emerson Process Management, 2005).

A.7.1.4 Instrument Air Quality

The current to pressure (I/P) and relay monitoring diagnostic can identify problems such as plugging in the I/P primary or in the I/P nozzle, instrument diaphragm failures, I/P instrument O-ring failures, and I/P calibration shifts. This diagnostic is particularly useful in identifying problems from contaminants in the air supply and from temperature extremes (Emerson Process Management, 2005).

A.7.1.5 In-Service Friction and Friction Trending

The in-service friction and dead band diagnostic determines friction in the valve assembly. Friction diagnostics data are collected and trended to detect valve changes that affect process control (Emerson Process Management, 2005).

A.7.1.6 Other Examples

In-service custom diagnostics can be configured to collect and graph any measured variable of a smart valve. Custom diagnostics can locate and discriminate faults not detectable by other means. Often, these faults are complicated and require outside expertise. In such cases, data are collected by local maintenance personnel and is then sent to an expert for further analysis, thus avoiding the costs and delays associated with an on-site visit.

In addition, online, in service safety checks, including partial stroke testing of safety valves, can be performed.

A.8 CONTROL VALVE MAINTENANCE

It is imperative to install block valves upstream and downstream of the control valve inside a bypass line that also includes a block valve to service control valves without shutting down the process. A drain or bleed valve should also be installed to relieve the pressure and purge the line before servicing the control valve.

The three control valve maintenance approaches are:
- Reactive—Action is taken after an event has occurred.
- Preventive—Action is taken on a schedule.
- Predictive—Action is taken based on diagnostic evaluation methods.

Reactive maintenance has several disadvantages that may affect reliability. Subtle deficiencies may go unnoticed and untreated simply because there is no clear indication of a problem. Critical valves might be neglected until they leak badly or fail to stroke. Maintenance is based on feedback from production, but valves might be removed unnecessarily on the suspicion of malfunction. Large valves or those welded in-line can require considerable time for removal, disassembly, inspection, and reinstallation. Time and resources could be wasted without solving the problem if the symptoms are actually caused by some other part of the system.

Preventive maintenance programs usually provide a significant improvement over reactive maintenance. Such programs result in servicing some valves that need no repair or adjustment and leaving others in the system long after they have stopped operating efficiently. Also, operations often extend the time between turnarounds to multiple years to maximize process availability. These extended run times offer less opportunity for traditional, out-of-service valve diagnostics.

The traditional maintenance process consists of several steps:

Fault Detection

Most of the valve maintenance effort is spent in monitoring valves while in service to detect the occurrence of a fault. When a fault is identified, the maintenance process transitions to fault discrimination.

Fault Discrimination

During this step, valve assets are evaluated to determine the cause of the fault and to establish a course of corrective action.

Process Recovery

Corrective action is taken to fix the source of the defect.

Validation

In the final step, valve assets are evaluated relative to either an "as new" condition or the last established baseline condition. Once validated, the maintenance process returns to fault detection status.

Some specific correct action and maintenance procedures are discussed in the sections below.

A.8.1 Actuator Diaphragm

Most pneumatic spring-and-diaphragm actuators use a molded diaphragm. The molded diaphragm facilitates installation, provides a relatively uniform effective area throughout valve travel, and permits greater travel than could be possible with a flat-sheet diaphragm. If a flat-sheet diaphragm is used for emergency repair, then replace it with a molded diaphragm as soon as possible (Emerson Process Management, 2005).

A.8.2 Stem Packing

Packing provides the pressure seal around the stem of a globe-style or angle-style valve body and should be replaced if leakage develops around the stem, or if the valve is completely disassembled for other maintenance or inspection.

Assure there is no pressure in the valve body before loosening packing nuts.

Remove the actuator and valve bonnet and pull out the stem. Push or drive the old packing out the top of the bonnet. Do not use the valve plug stem because the threads could sustain damage.

Clean the packing box and inspect the stem for scratches or imperfections that could damage new packing.

Check the trim and other parts. After reassembling, tighten body/bonnet bolting in a sequence similar to process recommended for flanges.

Slide the new packing parts over the stem in proper sequence and carefully so that the stem threads do not damage the packing rings. Adjust the packing by following the manufacturer's instructions (Emerson Process Management, 2005).

A.8.3 Seat Rings

Severe service conditions can damage the seating surface of the seat ring(s) so that the valve does not shut off satisfactorily. Grinding or lapping the seating surfaces will improve shutoff if damage is not severe. For severe damage, replace the seat ring (Emerson Process Management, 2005).

A.8.4 Grinding Metal Seats

The condition of the seating surfaces of the valve plug and seat ring can often be improved by grinding. Many grinding compounds are available commercially. For cage-style constructions, bolt the bonnet or bottom flange to the body with the gaskets in place to position the cage and seat ring properly and to help align the valve plug with the seat ring while grinding.

On double-port bodies, the top ring normally grinds faster than the bottom ring. Under these conditions, continue to use the grinding compound on the bottom ring, but use only a polishing compound on the top ring. If either of the ports continues to leak, use more grinding compound on the seat ring that is not leaking and polishing compound on the other ring. This procedure grinds down the seat ring that is not leaking until both seats touch at the same time. Never leave one seat ring dry while grinding the other.

After grinding, clean seating surfaces, and test for shutoff. Repeat grinding procedure if leakage is still excessive (Emerson Process Management, 2005).

A.8.5 Replacing Seat Rings

Follow the manufacturer's instructions. For threaded seat rings, use a seat ring puller. Before trying to remove the seat ring(s), check to see if the ring has been tack-welded to the valve body. If so, cut away the weld.

On double-port bodies, one of the seat rings is smaller than the other. On direct-acting valves install the smaller ring in the body port farther from the bonnet before installing the larger ring. On reverse-acting valves install the smaller ring in the body port closer to the bonnet before installing larger ring.

Remove all excess pipe compound after tightening the threaded seat ring. Spot weld a threaded seat ring in place to ensure that it does not loosen (Emerson Process Management, 2005).

A.8.6 Bench Set

Bench set is initial compression placed on the actuator spring with a spring adjuster. For air-to-open valves, the lower bench set determines the amount of seat load force available and the pressure required to begin valve-opening travel. For air-to-close valves, the lower bench set determines the pressure required to begin valve-closing travel. Seating force is determined by pressure applied minus bench set minus spring compression due to travel. Because of spring tolerances, there might be some variation in the spring angle. The bench set when the valve is seated requires the greatest accuracy (Emerson Process Management, 2005).

REFERENCES

Bishop, T., Chapeaux, M., Jaffer, L., Nair, K., Patel, S., November 2002. Ease control valve selection. Chemical Engineering Progress 98 (11).

Emerson Process Management, Control Valve Handbook, fourth ed., 2005. Fisher Controls International LLC.

GPSA Engineering Data Book, thirteenth ed., 2012. Gas Processors Suppliers Association (GPSA), Tulsa, OK, USA.

International Society of Automation, Recommended Practice 75.23, 1995. Considerations for Evaluating Control Valve Cavitation.

International Society of Automation, Recommended Practice 75.01, 2007. Flow Equations for Sizing Control Valves.

Rhinehart, N., March 2006. Rethink your control valve maintenance. Chem. Process. 69 (3).

Ruel, M.A., Simple Method to Determine Control Valve Performance and its Impacts on Control Loop Performance, http://www.topcontrol.com/fichiers/en/controlvalve performance.pdf.

APPENDIX B

Process Measurement and Instrumentation

B.1 INTRODUCTION TO MEASUREMENT

Measurement is the process by which one can convert physical parameters to meaningful numbers by comparing a quantity whose magnitude is unknown and a predefined standard. The result is expressed in numerical values. The measuring process is one in which the property of an object or system under consideration is compared with an accepted standard unit for that particular property.

B.2 MEASUREMENT SYSTEM APPLICATIONS

Measurement systems in natural gas processing are used primarily for monitoring and control of processes and operations. Certain applications of measuring instruments have a monitoring function only. They simply indicate the value or condition of a parameter (usually a physical quantity) and their readings do not serve any control functions. Other applications of measurements are in automatic control systems as well as monitoring. To control a parameter it must be measured.

The required accuracy of a measurement depends on the application. Accuracy is crucial for critical measurements that affect safety, reliability, and payments. Accuracy is not as crucial for trending and for many less critical control applications. However, repeatability is always important for any measurement application.

B.3 ELEMENTS OF A MEASUREMENT SYSTEM

Instrumentation in a natural gas processing plant usually comprises a system of pneumatic and electronic devices for measurement and control of all the process variables (pressure, flow, temperature, etc.) that are pertinent to the operation of the plant. In most cases, a computer is included in the instrumentation system to handle functions such as data gathering and transmission, bulk data storage, display, alarms, logging, computations, and control. The microprocessor has also led to many types of instruments becoming more intelligent or "computerized."

Instruments can be broadly classified into two categories: absolute and secondary. Absolute instruments give the magnitude of the quantity under measurement in terms of physical constants. Secondary instruments measure a quantity by observing the output indicated by the instrument. These instruments are calibrated by comparison with an absolute instrument.

In simple cases, the measuring system can consist of only a single unit that gives an output reading or signal according to the magnitude of the unknown variable applied to it. However, in more complex measurement situations, a measuring system consists of several separate elements as shown in Fig. B.1. (Beckwith, 1993).

Any instrument operation can be described in terms of functional elements.

- Primary sensing element

  The first element in any measuring system is the primary sensor, which gives an output that is a function of the measurand. For most but not all sensors, this function is at least approximately linear. Some examples of primary sensors are a thermocouple and a strain gauge.

- Variable conversion element

  Variable conversion elements are needed where the output variable of a primary transducer is in an inconvenient form and has to be converted to a more convenient form. For instance, the displacement measuring strain gauge has an output in the form of a varying resistance. The resistance change cannot be easily measured and so it is converted to a change in voltage by a bridge circuit, which is a typical example of a variable conversion element. In some cases, the primary sensor and variable conversion element are combined and the combination is known as a transducer.

- Variable processing element

  Signal processing elements exist to improve the quality of the output of a measurement system in some way. A very common type of signal processing element is the electronic amplifier, which magnifies the output of the primary transducer or variable conversion element, thus improving the sensitivity and resolution of measurement. This element of a measuring system is particularly important where the primary transducer has a low output. For example, thermocouples have a typical output of only a few millivolts. Other types of signal processing element are those that filter out induced noise and remove mean levels etc. In some devices, signal processing is incorporated into a transducer, which is then known as a transmitter.

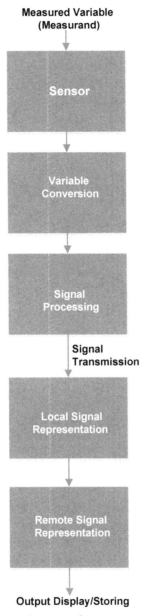

Figure B.1 Elements of a measuring system.

- Data presentation element

  The information about the quantity measured must be conveyed to the person handling the instrument or the system for monitoring, control, or analysis purposes. The information conveyed must be in a form intelligible to the personal or to the intelligent instrumentation system. This function is performed by a data presentation element. When data are monitored, continuous visual display is needed. These devices may be analog or digital indicating instruments.

  Data captured electronically can be recorded and historized.

B.4 COMMON MEASURED VARIABLES

Some typical, basic measured variables are:

- Flow
- Temperature
- Pressure
- Level
- Speed
- Vibration

  These are described in more detail in the following discussion.

B.4.1 Flow Rate

Flow measurement is critical to determine the amount of material purchased and sold, and in these applications, very accurate flow measurement is required. In addition, flows throughout the process should be regulated near their desired values with small variability; in these applications, good reproducibility is usually sufficient.

Flowing systems require energy, typically provided by pumps and compressors, to produce a pressure difference as the driving force, and flow sensors should introduce a small flow resistance, increasing the process energy consumption as little as possible. Most flow sensors require straight sections of piping before and after the sensor; this requirement places restrictions on acceptable process designs, which can be partially compensated by straightening vanes placed in the piping.

Several sensors rely on the pressure drop or head occurring as a fluid flows by a resistance. In these cases the relationship between flow rate and pressure difference is determined by the Bernoulli equation, assuming that changes in elevation, work, and heat transfer are negligible.

$$P_1/\rho + \upsilon_1^2/2g = P_2/\rho + \upsilon_2^2/2 + \sum f \qquad (B.1)$$

where D_i = diameter of the pipe; D_o = diameter of the orifice; υ_1 = upstream fluid velocity; υ_2 = fluid velocity through the orifice; P_1 = fluid upstream pressure; P_2 = fluid downstream pressure; ρ = fluid density and where $\sum f$ represents the total friction loss that is usually assumed negligible. This equation can be simplified and rearranged to give (Foust et al., 1980; Janna, 1993):

$$Q = C_m Y A_2 \sqrt{\frac{2(P_1 - P_2)}{\rho \left(1 - \frac{A_o^2}{A_i^2}\right)}} \tag{B.2}$$

where Q = volumetric flow rate; A_i = cross-sectional area of the pipe; A_o = cross-sectional area of the orifice hole.

The meter coefficient, C_m, accounts for all nonidealities, including friction losses, and depends on the type of meter, the ratio of cross-sectional areas and the Reynolds number. The compressibility factor, Y, accounts for the expansion of compressible gases; it is 1.0 for incompressible fluids. These two factors can be estimated from correlations (ASME, 1971; Janna, 1993) or can be determined through calibration. Eq. (B.2) is used for designing head flow meters for specific plant operating conditions.

When the process is operating, the meter parameters are fixed, and the pressure difference is measured. Then, the flow can be calculated from the meter equation, using the appropriate values for C_m and Y. All constants are combined, leading to the following relationship:

$$Q = C_o \sqrt{\frac{(P_1 - P_2)}{\rho_o}} \tag{B.3}$$

The equation can be simplified if the density is assumed its design value of ρ_o. This is a good assumption for liquid and can provide acceptable accuracy for gases in some situations. Again, all constants can be combined (including ρ_o) into C_1 to give the following relationship:

$$F = C_1 \sqrt{P_1 - P_2} \tag{B.4}$$

The flow is determined from Eq. (B.4) by taking the square root of the measured pressure difference, which can be measured by many methods. Typically, a diaphragm is used for measuring the pressure drop; a diaphragm with one pressure on each side will deform according to the pressure difference.

If the density of a gas varies significantly because of variation in temperature and pressure (but not average molecular weight), correction is usually based on the ideal gas law using low cost sensors to measure T and P according to:

$$F = C_1 \sqrt{\frac{(P_1 - P_2)}{\rho_o}} \sqrt{\frac{P_o T}{P T_o}} \tag{B.5}$$

where the density (assumed constant at ρ_o), temperature (T_o), and pressure (P_o) were the base case values used in determining C_o. If the density varies significantly due to composition changes and high accuracy is required, the real-time value of fluid density (ρ) can be measured by an on-stream analyzer for use as ρ_o in Eq. (B.3) (Clevett, 1986).

The difference between P_1 and P_2 is the nonrecoverable pressure drop of the meter that requires energy to overcome and increases the cost of plant operation.

The low pressure at the point of highest velocity creates the possibility for the liquid to partially vaporize (called flashing) or it might return to a liquid as the pressure increases after the lowest pressure point (called cavitation). We want to avoid any vaporization to ensure proper sensor operation and to retain the relationship between pressure difference and flow. Vaporization can be prevented by maintaining the inlet pressure sufficiently high and the inlet temperature sufficiently low.

Some typical head meters are described.

B.4.1.1 Orifice

An orifice plate is a restriction with an opening that is smaller than the pipe diameter. The typical orifice plate has a concentric, sharp-edged opening, as shown in Fig. B.2. Because of the smaller area, the fluid velocity increases to

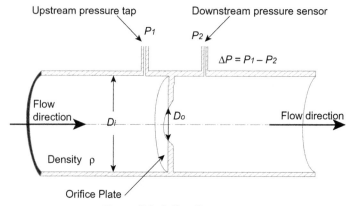

Figure B.2 Orifice flow meter.

cause a corresponding decrease in pressure. The flow rate can be calculated from the measured pressure drop across the orifice plate, $P_1 - P_2$. The orifice plate is the most commonly used flow sensor, but it creates a rather large nonrecoverable pressure due to the turbulence around the plate, leading to high energy consumption (Foust et al., 1980).

B.4.1.2 Venturi Tube

The Venturi tube shown in Fig. B.3 is similar to an orifice meter, but it is designed to form drag by nearly eliminating boundary layer separation. The change in cross-sectional area in the Venturi tube causes a pressure change between the convergent section and the throat, and the flow rate can be determined from this pressure drop. Although more expensive that an orifice plate, the Venturi tube introduces substantially lower non-recoverable pressure drops (Foust et al., 1980).

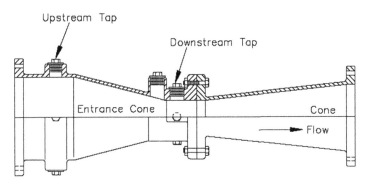

Figure B.3 Venturi flow meter.

B.4.1.3 Pitot Tube and Annubar

The pitot tube, shown in Fig. B.4, measures the static and dynamic pressures of the fluid at one point in the pipe. The flow rate can be determined from the difference between the static and dynamic pressures, which is the velocity head of the fluid flow. An annubar consists of several pitot tubes placed across a pipe to provide an approximation to the velocity profile, and the total flow can be determined based on the multiple measurements. Both the pitot tube and annubar contribute very small pressure drops, but they are not physically strong and should be used only with clean fluids.

The following flow sensors are based on physical principles other than head.

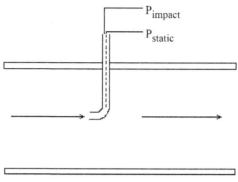

Figure B.4 Pitot flow meter.

B.4.1.4 Turbine Meters

As fluid flows through the turbine, it causes the turbine to rotate with an angular velocity that is proportional to the fluid flow rate. The frequency of rotation can be measured and used to determine flow. This sensor should not be used for slurries or systems experiencing large, rapid flow or pressure variation.

B.4.1.5 Vortex Shedding

Fluid vortices are formed against the body introduced in the pipe. These vortices are produced from the downstream face in an oscillatory manner. The shedding is sensed using a thermistor and the frequency of shedding is proportional to volumetric flow rate.

B.4.1.6 Positive Displacement

In these sensors, the fluid is separated into individual volumetric elements and the number of elements per unit time is measured. These sensors provide high accuracy over a large range. These meters are used to prove custody transfer meters (Table B.1).

B.4.2 Temperature

Resistance temperature detectors (RTD) and thermocouples are the most common devices used for measuring temperature.

B.4.2.1 Resistance Temperature Detectors

An RTD is a passive circuit element whose resistance is greater at higher temperature in a predictable manner. The traditional RTD element is constructed of a small coil of platinum, copper, or nickel wire, wound to a

Table B.1 Comparison of flow sensors

Sensor	Rangeability[1]	Accuracy[2]	Advantages	Disadvantages
Orifice	3.5:1	2–4% of full span	Low cost, well-documented calculations	High pressure loss
Venturi	3.5:1	1% of full span	Lower pressure loss than orifice, slurries do not plug	High cost, not applicable to pipe less than 6 inches
Annubar	3:1	0.5–1.5% of full span	Low pressure loss, good for large pipe diameters	Poor performance with dirty or sticky fluids
Turbine meter	20:1	0.25% of measurement	Wide rangeability, good accuracy	High cost, strainer needed, especially for slurries
Vortex shedding	10:1	1% of measurement	Wide rangeability, insensitive to variations in density, temperature, pressure, and viscosity	Expensive
Positive displacement	10:1 or greater	0.5% of measurement	High rangeability, good accuracy	High pressure drop, damaged by flow surge or solids

Inertial meters such as a Coriolis can directly measure mass flow rate.
[1]Rangeability is the ratio of full span to smallest flow that can be measured with sufficient accuracy.
[2]Accuracy applies to a calibrated instrument.

precise resistance value around a ceramic or glass bobbin. The winding is generally a helix style for industrial use.

The helix wire-wound RTD uses a bifilar wound coil wrapped around a bobbin and then sealed with molten glass, ceramic cement, or some other high-temperature insulating coating. The helix winding style helps protect the wire element from shock and vibration-induced changes to its resistance, but it may still be prone to stress-induced resistance change due to the different coefficients of thermal expansion of the wire coil and bobbin material.

The most common RTD element material is platinum, as it is a more accurate, reliable, chemically resistant, and stable material, making it less susceptible to environmental contamination and corrosion than the other metals. It is also easy to manufacture and widely standardized with readily available platinum wire available in very pure form with excellent reproducibility of its electrical characteristics. Platinum also has a higher melting point, giving it a wide operating temperature range (Table B.2).

Table B.2 Resistance temperature detectors element material usable temperature range

Platinum	$-260°C$ to $+650°C$ ($-436°F$ to $+1202°F$)
Nickel	$-100°C$ to $+300°C$ ($-148°F$ to $+572°F$)
Copper	$-75°C$ to $+150°C$ ($-103°F$ to $+302°F$)
Nickel/Iron	$0°C$ to $+200°C$ ($32°F$ to $+392°F$)

For an RTD sensor, it is the wires, which connect to the sensing element and the wire insulation, that generally limit the maximum application temperature of the sensor.

Measuring the temperature of an RTD involves measuring this resistance accurately. To measure the resistance, it is necessary to convert it to a voltage and use the voltage to drive a differential input amplifier. The use of a differential input amplifier is important as it will reject the common mode noise on the leads of the RTD and provide the greatest voltage sensitivity. The RTD signal is generally measured one of two ways: by connecting the RTD element in one leg of a Wheatstone bridge excited either by a constant reference voltage or by running it in series with a precision current reference and measuring the corresponding IR voltage drop. The latter method is generally preferred as it has less dependence on the reference resistance of the RTD element.

B.4.2.2 Thermocouples

A thermocouple comprises at least two metals joined together to form two junctions. One junction is connected to the body whose temperature is to be measured. The other junction is connected to a body of known temperature and is the reference junction. Therefore the thermocouple measures unknown temperature of the body with reference to the known temperature of the other body.

Appendix B: Process Measurement and Instrumentation   253

The working principle of thermocouple is based on three effects, discovered by Seebeck, Peltier, and Thomson. They are as follows:

1. Seebeck effect: The Seebeck effect states that when two different or un-like metals are joined together at two junctions, an electromotive force (emf) is generated at the two junctions. The amount of emf generated is different for different combinations of the metals.

2. Peltier effect: As per the Peltier effect, when two dissimilar metals are joined together to form two junctions, emf is generated within the circuit due to the different temperatures of the two junctions of the circuit.

3. Thomson effect: As per the Thomson effect, when two unlike metals are joined together forming two junctions, the potential exists within the circuit due to temperature gradient along the entire length of the conductors within the circuit.

In most of the cases, the emf suggested by the Thomson effect is very small and it can be neglected by making proper selection of the metals. The Peltier effect plays a prominent role in the working principle of the thermocouple.

The general circuit for the working of thermocouple comprises two dissimilar metals, A and B. These are joined together to form two junctions, p and q, which are maintained at the temperatures T1 and T2, respectively. Since the two junctions are maintained at different temperatures, the Peltier emf is generated within the circuit and is a function of the temperatures of the two junctions.

If the temperature of both the junctions is same, equal and opposite emf will be generated at both junctions and the net current flowing through the junction is zero. If the junctions are maintained at different temperatures, the emfs will not become zero and there will be a net current flowing through the circuit.

The device for measuring the current is connected within the circuit of the thermocouple. The output obtained from the thermocouple circuit is calibrated directly against the unknown temperature. Thus the voltage or current output obtained from thermocouple circuit gives the value of unknown temperature directly.

The amount of emf developed within the thermocouple circuit is very small, usually in millivolts; therefore, highly sensitive instruments should be used for measuring the emf generated in the thermocouple circuit. Two devices used commonly are the ordinary galvanometer and voltage balancing potentiometer. Of those two, a manually or automatically balancing potentiometer is used most often (Table B.3).

Table B.3 Standard thermocouple types and ranges

Type	Material	Normal Range (°C)
J	Iron-constantan	−190 to 760
T	Copper-constantan	−200 to 37
K	Chromel-alumel	−190 to 1260
E	Chromel-constantan	−100 to 1260
S	90% Platinum +10% rhodium-platinum	0−1482
R	87% Platinum +10% rhodium-platinum	0−1482

B.4.2.3 Choosing a Temperature Sensor

The choice of a temperature sensor should first consider:

- measurement range, including the range extensions of shutdown, startup, and process upset,
- response time, and
- stability, accuracy, and sensitivity in the application environment.

In general, if the application requires the highest accuracy, cost is not a concern, and the operating ambient temperature is less than 800°C, then the choice of an RTD over a thermocouple sensor is probably the right one. The RTD is more accurate, more stable, more repeatable, and offers a more robust output signal with better sensitivity and linearity than a thermocouple. However, the RTD does have a slower response time and a narrower operating range with a lower maximum operating temperature. An RTD is generally more expensive, and it does require excitation, which might drive the need for an external power source.

A thermocouple is of lower cost as well as simpler construction and has a wider temperature range and faster response time. Thermocouples also are available in many physical sizes and configurations. A type K thermocouple is the most common and least expensive of available thermocouple types with a wide operating temperature range with high sensitivity. It is constructed from nickel-based metals, which have good resistance to corrosion and are cheaper than the comparable platinum-based metals.

However, type K thermocouples have one lead that is magnetic and this might not work well around electric motors. It is also vulnerable to sulfur attack and should not be used in sulfurous atmospheres.

Although a thermocouple sensor system usually has a generally faster response time to changing temperature at its point of contact, it generally takes longer to reach thermal equilibrium after power is applied. If the distance between the measuring instrument and the sensor is very long, this might favor the use of a thermocouple over that of an RTD, because minimizing lead-wire resistance can be a big factor in long-distance RTD applications.

Whatever the material and construction, an RTD is still relatively bulky by comparison to the thermocouple, and as a result, cannot be used to measure temperature at a single point, like that of a thermocouple. On the other hand, RTD technology provides an excellent means of measuring the average temperature over a surface by spreading the resistance wire over that surface, and this ability to average temperature over a surface area (or an immersion depth) will be more desirable for some applications.

The use of a platinum RTD sensor may be preferred when a temperature measurement accuracy of better than 2°F or 1°C is required. By comparison, thermocouple accuracy will be about 2–4°C, typical. However, for point-of-contact measurement at temperatures above about 800°C (the maximum temperature at which platinum RTDs can be used), a thermocouple is the better choice due to its higher rated operating temperature.

B.4.3 Pressure

Most liquid and all gaseous materials in the process industries are contained within closed vessels. For the safety of plant personnel and protection of the vessel, pressure in the vessel is controlled. In addition, pressured is controlled because it influences key process operations like vapor–liquid equilibrium, chemical reaction rate, and fluid flow.

The following pressure sensors are based on deformation caused by force:
- Bourdon tube
- Bellows
- Diaphragm

A bourdon tube is a curved, hollow tube with the process pressure applied to the fluid in the tube. The pressure in the tube causes the tube to deform or uncoil. The pressure can be determined from the mechanical displacement of the pointer connected to the bourdon tube. Typical shapes for the tube are "C" (normally for local display), spiral, and helical.

A bellows is a closed vessel with sides that can expand and contract, like an accordion. The bellows itself or a spring can determine the position of the bellows without pressure. The pressure is applied to the face of the bellows, and its deformation and its position depend on the pressure.

A diaphragm is typically constructed of two flexible disks, and when a pressure is applied to one face of the diaphragm, the position of the disk face changes due to deformation. The position can be related to pressure.

The following pressure sensors are based on electrical principles; some convert a deformation to a change in electrical property, others a force to an electrical property:

- Capacitive or inductance
- Resistive, strain gauge
- Piezoelectric

The movement associated with one of the mechanical sensors can be used to influence an electrical property such as capacitance affecting a measured signal. For example, under changing pressure a diaphragm causes a change in capacitance or inductance.

The electrical resistance of a metal wire depends on the strain applied to the wire. Deflection of the diaphragm due to the applied pressure causes strain in the wire, and the electrical resistance can be measured and related to pressure.

A piezoelectric material, such as quartz, generates a voltage output when pressure is applied on it. Process pressure applies a force by the diaphragm to a quartz crystal disk that is deflected (Table B.4).

B.4.4 Liquid Level

The level of a liquid in a vessel should be maintained above the exit pipe. In addition, the level should not overflow an open vessel nor should it exit through a vapor line of a closed vessel, which could disturb a process designed for vapor.

Level is usually reported as a percentage of span, rather than in height. Sensors can be located in the vessel holding the liquid or in an external "leg," which acts as a manometer. When in the vessel, float and displacement sensors are usually placed in a "stilling chamber," which reduces the effects of flows in the vessel.

Some typical types of level measurement are:

- Float
- Displacement
- Differential pressure
- Capacitance

A float follows the movement of the liquid level. The material of construction of a float of material must be lighter than the fluid in the

Table B.4 Comparison of pressure sensors

Sensor	Limits of application	Accuracy	Dynamics	Advantages	Disadvantages
Bourdon, "C"	Up to 100 MPa	1–5% of full span	Slow	Low cost with reasonable accuracy, wide limits of application	Hysteresis, affected by shock and vibration
Spiral	Up to 100 MPa	0.5% of full span	Slow		
Helical	Up to 100 MPa	0.5–1% of full span	Slow		
Bellows	Typically vacuum to 500 kPa	0.5% of full span	Slow	Low cost, differential pressure, good for vacuum	Smaller pressure range of application, temperature compensation needed
Diaphragm	Up to 60 kPa	0.5–1.5% of full span	Slow	Very small span possible	Usually limited to low pressures below 8 kPa
Capacitance/ inductance	Up to 30 kPa	0.2% of full span	Slow	Accuracy	Low range of pressure
Resistive/ strain gauge	Up to 100 MPa	0.1–1% of full span	Fast	Large range of pressures	Susceptible to electromagnetic interference
Piezoelectric	Any	0.5% of full span	Very fast	Fast dynamics, very small	Sensitive to temperature changes

vessel. The position of the float, usually attached to a rod, can be determined to measure the level.

By Archimedes principle, a body immersed in a liquid is buoyed by a force equal to the weight of the liquid displaced by the body. Thus a body that is denser than the liquid can be placed in the vessel and the amount of liquid displaced by the body, measured by the weight of the body when in the liquid, can be used to determine the level.

The difference in pressures between two points in a vessel depends on the fluid between these two points. If the difference in densities between the fluids is significant, which is certainly true for a vapor and liquid and can be true for two different liquids, the difference in pressure can be used to determine the interface level between the fluids. Usually, a seal liquid is used in the two connecting pipes (legs) to prevent plugging at the sensing points.

A capacitance probe can be immersed in the liquid of the tank, and the capacitance between the probe and the vessel wall depends on the level. By measuring the capacitance of the liquid, the level of the tank can be determined (Table B.5).

Table B.5 Comparison of level sensors

Sensor	Limits of application	Advantages	Disadvantages
Float	Up to 1 m	Can be used for switches	Cannot be used with sticky fluids that coat the float
Displacement	0.3—3 m	Good accuracy	Limited range, cost of external mounting for high pressures
Differential pressure	Essentially no upper limit	Good accuracy, large range, applicable to slurries with use of sealed lines	Assumes constant density, sealed lines sensitive to temperature
Capacitance	Up to 30 m	Applicable for slurries, level switch for many difficult fluids	Affected by density variations

B.4.5 Speed

Rotating machinery requires speed sensing for functional systems such as control, monitoring, and safety. Speed sensing is accomplished using primarily passive magnetic pickups (MPUs) or active proximity probes (Ried, 2011).

MPUs are commonly selected because of simplicity, reliability, and low cost. MPUs are passive probes in the sense that there are no active signal conditioning electronics in the probe. Passive probes have variable signal outputs that depend on many factors and can be difficult to apply.

Sensors with active signal conditioning integrated into the sensor are classified as active probes. Active probes usually have more consistent signal output because of the integrated electronics, but with the associated increase in complexity.

For industrial applications, API 670 has guidelines on the application of speed sensors. Specifically, Annex J contains well-proven information on relative dimensions of the MPU and speed sensing tooth parameters. Functional safety compliance with International Electrotechnical Commission (IEC) 61508 and IEC 61511 is increasingly required in new and retrofit systems. Passive probes are a good choice in these systems because they are type A (passive) devices per IEC 61508 and take up less of the safety allocation than active probes.

Passive probes cause a magnetic flux to be generated in the probe that is converted to an electrical signal by a coil inside the probe. The amplitude of this signal is proportional to:

- the amount of magnetic flux generated by the passing tooth,
- the number of turns in the coil, and
- loading on the coil due to the sensing circuit.

The amount of magnetic flux generated is a complex function of the probe pole face dimensions, the speed tooth dimensions, tooth speed, MPU and speed tooth magnetic properties, and the gap between the probe and the gear tooth. The physics behind the flux generation comes from one of Maxwell's equations that govern electromagnetic behavior. In a simplified form, it states that:

$$V = N \times d\varphi/dt \qquad (B.6)$$

where V = voltage; N = number of turns on the coil; $d\varphi/dt$ = the amount of magnetic flux generated by the passing tooth.

The flux change results from a change in the magnetic coupling as the tooth edge passes the pole face. The amount of flux in the magnetic circuit also depends on the material properties of the MPU core and the tooth. The tooth must be a ferromagnetic material like annealed iron and not paramagnetic like aluminum or stainless steel.

B.4.6 Vibration

Sensors used to measure vibration come in three basic types: displacement, velocity, and acceleration. Displacement sensors measure changes in distance between a machine's rotating element and its stationary housing (frame). Displacement sensors come in the form of a probe that threads into a hole drilled and tapped in the machine's frame, just above the surface of a rotating shaft. Velocity and acceleration sensors measure the velocity or acceleration of the element attached to the sensor, which is usually some external part of the machine frame.

Some displacement sensor designs use magnetic eddy current technology to sense the distance between the probe tip and the rotating machine shaft. The sensor itself is an encapsulated coil of wire, energized with high-frequency alternating current. The magnetic field produced by the coil induces eddy currents in the metal shaft of the machine, as though the metal piece were a short-circuited secondary coil of a transformer with the probe's coil as the transformer's primary winding. The closer the shaft moves toward the sensor tip, the tighter the magnetic coupling between the shaft and the sensor coil, and the stronger the eddy currents.

The oscillator circuit providing sensor coil excitation is called a proximitor. The proximitor module is powered by an external DC power source and drives the sensor coil through a coaxial cable. Proximity to the metal shaft is represented by a DC voltage output from the proximitor module, with 200 mV per mil of motion being the standard calibration.

Since the proximitor's output voltage is a direct representation of the distance between the probe's tip and the shaft's surface, a "quiet" signal (no vibration) will be a pure DC voltage. A technician adjusts the probe such that this quiescent voltage will lie between the proximitor's output voltage range limits. Any vibration of the shaft will cause the proximitor's output voltage to vary in precise step. The shaft vibration in cycles per second will cause the proximitor output signal to be the same waveform in hertz superimposed on the DC bias voltage set by the initial probe to shaft gap.

An oscilloscope connected to this output signal will show a direct representation of shaft vibration, as measured in the axis of the probe. It is customary to arrange a set of three displacement probes at the end of a machine shaft to measure vibration: two radial probes and one axial (or thrust) probe. The purpose of this triaxial probe configuration is to measure shaft vibration (and/or shaft displacement) in all three dimensions.

It is also common to see one phase reference probe installed on the machine shaft, positioned in such a way that it detects the periodic passing of a keyway or other irregular feature on the shaft. The "keyphasor" signal will consist of one large pulse per revolution.

The purpose of a keyphasor signal is twofold: to provide a reference point in the machine's rotation to correlate other vibration signals against and to provide a simple means of measuring shaft speed. The location in time of the pulse represents shaft position, whereas the frequency of that pulse signal represents shaft speed.

B.5 INSTRUMENT TYPES AND PERFORMANCE CHARACTERISTICS

B.5.1 Review of Instrument Types

There are two main types of the measuring instruments: analog and digital. Analog instruments indicate the magnitude of the quantity in the form of the pointer movement and the value is read according to markings on a scale. One has to learn reading such instruments since there are certain markings on the scale. The readings may not always be entirely correct, since some human error is always involved in interpretation.

Digital measuring instruments indicate the values of the quantity in digital format that is in numbers, which can be read easily. Since there is no human error involved in reading these instruments, they are more accurate than the analogue measuring instruments. Some digital measuring instruments can also carry out a number of calculations including highly complex equations.

The main advantages of digital instruments are:
- They are very easy to read.
- Since there are very few moving parts in the electronic instruments, they are usually more accurate than the analog instruments.
- The electronic items tend to be less expensive than the mechanical items.
- The output can be transmitted to a computer and can be recorded for future reference.
- The output of the digital devices can be obtained in the computer.
  There are also come disadvantages of the digital instruments, namely:
- Sometimes they tend to indicate erratic values due to faulty electronic circuit or damaged display.

- In case of high humidity and corrosive atmospheres the internal parts may become damaged and indicate faulty values.
- Sometimes these instruments show a reading even though there is no applied measurable parameter.

The advantages of digital instruments tend to outweigh the disadvantages, which is why they have become highly popular.

B.5.2 Necessity for Calibration

B.5.2.1 Errors During the Measurement Process

When any physical quantity is measured, we tend to assume that the instruments are accurate. There are three types of errors in instruments: assembly errors, environmental errors, and random errors (Jain, 2011).

Assembly errors are mistakes in the measuring instrument due to improper manufacturing of the instruments. Various components of the instrument are made separately and then they are assembled together. There may be errors in the individual components or in the assembly of the components. If there are assembly errors in the instruments, the instrument will just not give the correct reading and the user can hardly do anything about it, but return it and get a new one. Following are some typical assembly errors:

- Displaced scale—This is the incorrect fitting of the measuring scale. For instance, the zero of pointer may not coincide with actual zero on the scale. Sometimes the scale is cracked, thus showing faulty readings.
- Nonuniform scale—Sometimes the scale of the measuring instrument is not divided uniformly. In some part of the scale, the markings may be too close and in other parts too far.
- Bent pointer—The pointer may be bent in a horizontal or vertical direction. In either case, an erroneous reading is given.
- Manufacturing errors in the components—Instruments consist of a number of small components, which may be manufactured in different places. Sometimes there are manufacturing errors in some of these components such as gears, levers, links, hinges, etc.

Measuring instruments are assembled and calibrated in certain environmental conditions and are designed to be used within certain restricted conditions. When they are used in different conditions, there are errors in measurement, which are considered environmental errors. Most instruments are designed to be used within certain limits of temperature, pressure, humidity, altitude, etc. and when the limits are extended, there may be errors in the measurements.

It is much easier to find the assembly errors in instruments, but the errors due to change in environmental conditions are highly unpredictable. These errors can lead to high or low readings.

Some precautions to reduce the environmental errors in the instruments:

- Use instruments within the specified limits of temperature, pressure, and humidity for which the instrument has been designed. These limits are given in the instructions manual.
- If the instrument must be used beyond the specified limits of environmental conditions, then apply suitable corrections to the recorded measurement.
- Calibrate the instrument for the correct environmental conditions.
- Select a device that enables applying the compensation automatically.

Apart from assembly and environmental errors there can be random errors that may be very difficult to trace and predict. It is not possible to list all the possible random errors, but some of the most prevalent are described:

- Frictional errors—There are a number of moving mechanical parts in analog instruments. The friction between these components leads to error. Due to friction some of the parts wear, which further adds to the error of the instrument. Analog instruments should be replaced periodically.
- Mechanical vibrations—When the instrument is subject to vibrations, then the parts of the instrument start vibrating, giving faulty readings.
- Electrical interference—Electromagnetic effects may cause errors in digital instruments.
- Backlash in the movement—This is the error due to time lag between the application of the parameter and the instrument actually showing the reading.
- Hysteresis of the elastic members—Over time the elastic members tend to lose some elasticity leading to errors in the indicated value of the instrument.

B.5.3 Calibration of Measuring Sensors and Instruments

Instruments must be calibrated against the standard values of the physical quantities. Standard instruments, which are used very rarely for the sole purposes of calibration are kept in highly controlled air conditioned atmosphere so that their scale does not change with the external atmospheric changes.

Calibration of the measuring instrument is the process in which the readings obtained from the instrument are compared with the standards in the laboratory at several points along the scale of the instrument. As per the results obtained from the instrument readings and the standards, the curve is plotted. If the instrument is accurate there will be matching of the scales of the instrument and the standard. If there is deviation of the measured value from the instrument against the standard value, the instrument is calibrated to give the correct values.

All new instruments must be calibrated against some standard. After continuous use of the instrument for long periods, sometimes it loses its calibration. In such cases, the instrument can be calibrated again if it is in good reusable condition.

Even if the instruments in the factory are working in the good condition, it is always advisable to calibrate them at the regular intervals as specified by the manufacturer to avoid erroneous readings of highly critical parameters. This is very important especially where measurements are safety related.

B.6 ANALYZERS

Process analyzers are online tools used to determine the chemical composition. They enable process optimization, asset protection, and compliance with environmental regulations.

Important chemical compositions in natural gas processing operations include:
- Hydrocarbons
- Hydrogen sulfide
- Carbon dioxide
- Nitrogen
- Helium
- Water
- Mercury
- Oxygen
- Mercaptans

Gas chromatographs are probably the most widely used online analytical instruments in natural gas processing plants. These analyzers are effective for measuring a wide range of hydrocarbons, helium, oxygen, nitrogen, carbon dioxide, and hydrogen sulfide. Liquid streams are analyzed by vaporizing the sample. Several standards are available including:
- GPA 2261—Analysis of natural gas and similar gaseous mixtures by gas chromatography

- ASTM D1945—Standard test method for analysis of natural gas by gas chromatography
- ISO 6974—Natural gas: Determination of composition and associated uncertainty by gas chromatography

The main disadvantages of gas chromatographs are that the measurement is not instantaneous; they are not effective for trace components and are maintenance intensive. If the analysis of specific components is required continuously or at very low concentrations, then other methods may be preferred.

Tunable diode laser absorption spectroscopy (TDLAS) is gaining popularity for a wide variety of analyses. However, TDLAS is not very effective for mixtures of hydrocarbons.

The following summarizes typical techniques for online measurement of nonhydrocarbon components.

In addition to inlet gas, it is important to measure carbon dioxide content of treated gas, acid gases, sales gas, and natural gas liquids (NGL) product. Gas chromatography is effective for measuring concentrations of carbon dioxide in the percentage range. For lower concentrations or more frequent measurement, carbon dioxide is often measured with infrared techniques, whereas ultraviolet absorption spectroscopy and TDLAS are sometimes used depending on the application.

It is important to measure various sulfur species in many streams including inlet gas, treated gas, acid gases, sulfur plant tail gas, sales gas, and NGL product. Hydrogen sulfide, carbon disulfide, and carbonyl sulfide are common sulfur species that are not desirable in products, whereas mercaptans are injected into products purposely.

Colorimetric principles were the first techniques used for measuring hydrogen sulfide in process streams. The advantages are:

- Interference free (selective to hydrogen sulfide only),
- Detection limit <1 ppb,
- No zero drift,
- Five seconds response on breakthrough,
- Linear response, and
- No other utilities than power supply required for ranges up to 100 ppm.

In cases where other sulfur species in addition to hydrogen sulfide are measured, ultraviolet absorption spectroscopy is often used.

For atmospheric hydrogen sulfide detection, electrochemical and solid state sensors are used.

Advantages to using an electrochemical sensor include allowing sensors to be specific to a particular gas in the parts-per-million range, low power requirements, high accuracy, and lower cost than other gas detection technologies. However, electrochemical sensors are sensitive to temperature, subject to interference from other gases, and have a shorter lifespan with greater exposure to the targeted gas.

Solid-state sensors have the ability to detect both low parts-per-million levels of gases, as well as high combustible levels. However, they are not very specific to a particular gas as often required for industrial applications and could potentially provide false alarms. Solid-state sensors may be affected by fugitive volatile organic compounds and must be located in areas where these gases are not present.

Mercury detection in process streams is important when aluminum brazed heat exchangers are employed. Atomic fluorescence allows absolute detection levels of below 0.1 pg.

Moisture content is important in dehydrated gas streams. Quartz crystal analyzers are often used in this service.

Oxygen measurement is desirable in various services with flue gases the most common. Zirconium oxide cells are the most popular with TDLAS sometimes used.

Combustibles measurement is important in flue gas streams and for detection in enclosed spaces. Infrared and catalytic gas detectors are the most popular methods used to measure combustibles. Infrared is a good choice for flue gas streams and catalytic gas detectors in atmospheric conditions where hydrogen may be detected as well.

Online calorimeters are used to quickly and accurately measure the calorific value of gas streams. These analyzers use a combustion chamber or measure thermal conductivity and various temperatures. In combustion chamber devices, either the exhaust temperature can be controlled or the residual oxygen measured. These analyzers typically include the capability to measure specific gravity and therefore calculate the Wobbe Index.

Sampling systems for ex situ analyzers and placement of in situ systems is extremely important for accurate and reliable results.

B.7 PROCESS CONTROLLERS

Although some analog controllers still exist, the majority of basic process control is accomplished digitally with distributed control systems (DCSs), programmable logic controllers (PLCs), and sometimes programmable automation controllers.

B.7.1 Distributed Control Systems

DCSs derive from direct digital control and use decentralized elements or subsystems to control distributed processes or manufacturing systems. They offer flexibility, extended equipment life, simplicity of new equipment integration, and centralized maintenance when used in an industrial environment.

The architecture is based on digital communication between distributed controllers, workstations, and other computing elements. DCS backbone networks are typically standard Ethernet hardware but use their own closed, high-performance protocols and natively support redundancy. In DCS, the process networks are connected to controllers, which are connected to the process DCS backbone.

Many process control systems require redundancy for input/output (I/O), controllers, networks, and human–machine interface (HMI) servers at various levels. A DCS generally has easier-to-apply redundancy solutions but the open networking standard groups have defined solutions for their protocols particularly with the initiatives for networked machine safety.

A DCS provides multidisciplined controllers for logic, sequential and process control, HMIs, custom applications, and business integration on one platform. The controllers are networked to a central console and aim to centralize plant operations to allow control, monitoring, and reporting of individual components and processes at a single location.

A DCS is a hierarchical system, although not all systems share an identical hierarchy. Individual controllers, supervised by master controllers, make up the lowest level of the hierarchy. The master controllers connect to individual computers and servers, which are further connected to video output devices and HMI, which is the actual point of user control.

Many DCS components can also operate as standalone devices. Although a DCS ultimately governs the functionality of its networked components, the same components can often be reprogrammed for use in other applications.

Most DCSs are designed with redundant elements. Redundant engineering increases a system's reliability by using backup processors in case of primary processor failure. Redundant elements are necessary in DCS because safety critical processes are controlled and redundancy increases equipment reliability, leaving the DCS operator to concentrate on displays, software, and applications.

B.7.2 Programmable Logic Controllers

A PLC is an industrial solid-state computer that monitors inputs and outputs, and makes logic-based decisions for automated processes or machines. PLCs were introduced in the late 1960s to provide the same functions as relay logic systems. Relay systems at the time tended to fail and create delays. Relays switching under load can cause undesired arcing between contacts. Arcing generates high temperatures that weld contacts shut and cause degradation of the contacts in the relays, resulting in device failure. Technicians then had to troubleshoot an entire wall of relays to fix the problem. Replacing relays with PLCs helps prevent overheating of contacts.

PLCs are robust and can survive harsh conditions including severe heat, cold, dust, and extreme moisture. Their programming language is easily understood, so they can be programmed without much difficulty. PLCs are modular so they can be plugged into various setups.

PLC systems use open published protocols that are designed to cover a wide range of applications including simple discrete, synchronized motion control, motor control, and process. The level of controller redundancy for higher-level process applications is new to PLC suppliers and they continue to add options for redundancy.

PLCs do have disadvantages. They do not perform well when handling complex data. When dealing with data that requires C++, Visual Basic, or other sophisticated programming languages, computers are the controllers of choice. PLCs also cannot display data well, so external HMIs are often required.

A central processing unit (CPU) serves as the brain of the PLC. It is a 16- or 32-bit microprocessor consisting of a memory chip and integrated circuits for control logic, monitoring, and communicating. The CPU directs the PLC to execute control instructions, communicate with other devices, carry out logic and arithmetic operations, and perform internal diagnostics. The CPU runs memory routines, constantly checking the PLC to avoid programming errors and ensure the memory is undamaged.

Memory provides permanent storage to the operating system for data used by the CPU. The system's read-only memory stores data permanently for the operating system and random access memory stores status information for input and output devices, along with values for timers, counters, and internal devices. PLCs require a programming device, either a computer or console, to upload data onto the CPU.

PLCs read signals from different sensors and input devices. These input devices can be keyboards, switches, or sensors. Inputs can be either in digital or analog form. Robots and visual systems are intelligent devices that can send signals to PLC input modules. Output devices such as motors and solenoid valves complete the automated system.

Several programming languages are used in PLCs and are defined by the international standard IEC 61131. Ladder logic is one of the most commonly used PLC languages where symbols represent opening and closing relays, counters, timers, shift registers, and mathematical operations. The symbols are arranged into the desired program routine. Rules in ladder logic are termed "rungs." Each rung has a single output, but a single input can be found in more than one rung.

Another programming language is function block diagram (FBD). The function, represented by blocks, connects input and output variables. FBD is useful in depicting algorithms and logic from interconnected controls systems.

Structured Text (ST) is a high-level language that uses sentence commands. In ST, programmers can use "if/then/else," "SQRT," or "repeat/until" statements to create programs.

Instruction list is a low-level language with functions and variables defined by a simple list. Program control is done by jump instructions and subroutines with optional parameters.

Sequential Function Chart (SFC) language is a method of programming complex control systems. It uses basic building blocks that run their own subroutines. Program files are written in other programming languages. SFC divides large and complicated programming tasks into smaller and more manageable tasks.

RS-232, a serial communication standard that uses binary code to transmit data in American Standard Code of Information Interchange (ASCII) format, is the most common method PLCs use to communicate with external devices. ASCII translates letters and numbers into binary code that computers can read. ASCII is a 7-bit code (a bit being "1" or "0") that, when translated, results in 128 characters. PLC serial ports transmit and receive data as voltages. PLCs are byte oriented, so the eighth (or "parity byte") checks to see if data has been corrupted.

B.7.3 Human—Machine Interfaces

The HMI was developed as the window into the operation of an industrial process system. HMIs were once used to just display the active and inactive outputs in an industrial process and where external buttons were used as the input devices for the industrial process. With the advancement in interactive screens the HMI can be programmed for the user to not only visually inspect the outputs of the system but also use the HMI screen as an input to the system, forgoing the use of external buttons.

The user interface is the part of the machine that handles the human—machine interaction. Mice, keypads, and touchscreens are examples of the physical part of the HMI. The goal of this interaction is to allow effective operation and control of the machine from the human end while the machine simultaneously feeds back information that aids the operators' decision-making process.

The goal of user interface design is to produce a user interface that makes it easy and efficient to operate a machine in the way that produces the desired result. This means that the operator needs to provide minimal input to achieve the desired output and also that the machine minimizes undesired outputs to the human.

With the increased use of personal computers and the relative decline in societal awareness of heavy machinery, the term user interface is generally assumed to mean the graphical user interface, whereas industrial control panel and machinery control design discussions more commonly refer to HMIs.

Graphical user interfaces may be object, application, or web oriented. Newer implementations utilize PHP, Java, JavaScript, AJAX, Apache Flex, .NET Framework or similar technologies to provide real-time control in a separate program, eliminating the need to refresh a traditional HTML-based web browser.

B.7.4 Supervisory Control and Data Acquisition

Supervisory control and data acquisition (SCADA) is used for acquiring data from remote devices such as valves, pumps, transmitters, etc. and providing overall control remotely from an SCADA Host software platform. This provides process control locally so that these devices turn on and off at the right time, supporting a remote method of control and monitoring of these processes. SCADA Host platforms also provide functions for graphical displays, alarming, trending, and historical storage of data.

SCADA provides solutions for wide area networks that rely on tenuous communication links. These types of SCADA systems are used extensively throughout the natural gas industry primarily for operating gathering systems and pipelines because assets are spread over large geographical areas.

The overall structure of an SCADA system includes:

- Field instrumentation,
- PLCs and remote telemetry units (RTUs),
- Communications networks, and
- SCADA Host software.

Originally, RTUs just included well-developed communication interfaces or telemetry. In the 1990s control programming was added to the RTU so it operated more like a PLC. A further development of devices in the field is to offer a specific application that incorporates a number of instruments and devices with an RTU. Technology sets are provided for an "off-the-shelf" approach to common process requirements, e.g., gas well production that includes elements of monitoring, flow measurement and control that would extend as an asset into the SCADA Host. RTUs are designed to operate in extremely harsh environments.

The remote communication network is necessary to relay data from RTUs, which are out in the field or along the pipeline, to the SCADA host located at the field office or central control center. Communications are typically radio, cellular, or satellite communications. Protocols for exchanging data between SCADA and RTUs or PLCs are often a combination of a standard such as MODBUS and proprietary elements to meet specific functionality requirements. In recent years, protocols have appeared that are truly nonproprietary, such as the Distributed Network Protocol.

Traditionally, SCADA Host software has been the mechanism to view graphical displays, alarms, and trends. Control from the SCADA Host became available when control elements for remote instruments were developed. With advancements in Information Technology (IT) there is a drive for the SCADA Host to be an Enterprise entity providing data to a number of different users and processes. This has encouraged SCADA Host software development to adopt standards and mechanisms to support interfacing to these systems. It also means that IT, traditionally separated from SCADA systems, is now involved in helping to maintain networks, database interfacing, and user access to data.

The SCADA Host uses integrated software programs called "drivers." Software drivers contain the different types of protocols to communicate with remote devices such as RTUs and PLCs. SCADA Host software platforms also include integral databases specifically designed for SCADA Host software requirements, being able to handle thousands of changes a second, for really large systems, yet still conform to standard database interfacing. These standards are required so that third-party databases can access data from the SCADA Host software. Remote client access to the SCADA Host enables users to operate and monitor SCADA systems while on the move between or at other locations.

There is a drive toward operational safety for SCADA Host systems within the oil and gas industry. Regulations such as 49 CFR 195.446 Control Room Management address SCADA Host software and how it functions in terms of operations, maintenance, and management. It also covers the degree of integration of the SCADA system itself and its use of open architecture and standards.

B.7.5 Security

Security for control systems has become an important topic. Traditionally control systems were isolated entities that were the realm of operators, engineers, and technicians. This has meant that control platforms were not necessarily developed to have protected connections to public networks. This left many control platforms open to attack, as they did not have the tools necessary to protect themselves.

There are a number of standards available that describe how to secure a control system, not just in terms of the technology employed, but in terms of practices and procedures. These practices and procedures include items of training, access and procedures to follow when security has been compromised.

B.8 SAFETY INSTRUMENTATION

All instrumentation should be designed with safety in mind. Automated systems should be configured to fail to a safe condition. Other systems such as pressure relief are designed to prevent conditions that would lead to equipment damage and containment of materials.

B.8.1 Relief Valves and Relief Systems

A relief system is designed to discharge material during abnormal conditions to the atmosphere to relieve pressures in excess of the maximum allowable working pressure (MAWP) of containment vessels.

The relief system may include:
- The relief device
- The collection piping
- Flashback protection
- A gas outlet

In addition, a scrubbing vessel should be provided for liquid separation if liquid hydrocarbons are anticipated.

There are three main engineering considerations when designing or modifying a relief system:

- Determining the relief requirements of individual pieces of equipment and selecting the appropriate devices to handle the imposed loads.
- Designing a relief header system that will handle the imposed loads or expansion modifications.
- Defining reasonable total relief loads for the combined relief header or disposal system and designing an appropriate disposal system with minimum adverse impact to personnel safety, plant process system integrity, and the environment.

There are a number of industry codes, standards, and recommended practices that provide guidance in the sizing, selection, and installation of relief devices and systems. The American Society of Mechanical Engineers (ASME) Pressure Vessel Code, Sec. VIII, Division 1, paragraph UG-127, lists the relief-valve code requirements (ASME, 2015). RP 520, Part 1, provides an overview of the types of relief devices, causes of overpressure, relief-load determination, and procedures for selecting and sizing relief devices (API, 2014a). RP 520, Part 2, provides guidance on the installation of relief devices (API, 2015), and RP 521 provides guidance on the selection and design of disposal systems (API, 2014b).

The most common causes of overpressure in upstream operations are blocked discharge, gas blow by, and fire. When the worst-case relief load is caused by a control valve failing to open, the relief device should be sized with full-sized trim in the control valve, even if the actual valve has reduced trim. When the worst-case relief load is caused by gas blow by, the relief device should be sized with full-sized trim in the smallest valve in the liquid outlet line, even if the actual valve has reduced trim.

Many vessels are insulated for energy savings. Thermal insulation limits the heat absorption from fire exposure as long as it is intact. It is essential that effective weather protection be provided so that insulation will not be removed by high-velocity fire hose streams.

The two primary types of relief devices are the relief valve and rupture disk.

The three basic types of pressure relief valves are conventional spring loaded, balanced spring loaded, and the pilot operated.

In the conventional spring-loaded valve, the bonnet, spring, and guide are exposed to the released fluids. If the bonnet is vented to the atmosphere, relief-system backpressure decreases the set pressure. If the bonnet is vented internally to the outlet, relief-system backpressure increases the set pressure. The conventional spring–loaded valve is used in noncorrosive services and where backpressure is less than 10% of the set point.

The balanced spring–loaded valve incorporates a means to protect the bonnet, spring, and guide from the released fluids and minimizes the effects of backpressure. The disk area vented to the atmosphere is exactly equal to the disk area exposed to backpressure. These valves can be used in corrosive or dirty service and with variable backpressure.

The pilot-operated valve is combined with and controlled by an auxiliary pressure pilot. The resistance force on the piston in the main valve is assisted by the process pressure through an orifice. The net seating force on the piston actually increases as the process pressure nears the set point.

The rupture disk device is a non-reclosing differential pressure device actuated by inlet static pressure. The rupture disk is designed to burst at set inlet pressure. The device includes a rupture disk and a disk holder. The rupture disk may be used alone, in parallel with, or in conjunction with pressure relief valves. They are manufactured in a variety of materials with various coatings for corrosion resistance.

The entire relief system must be considered before selecting the appropriate relief device. The relief headers should be designed to minimize pressure drop, thus allowing for future expansion and additional relief loads.

Relief devices are normally set to relieve at the MAWP. The greater the margin between the set pressure and the operating pressure, the less likelihood there is of leakage. Aside from the requirements to compensate for superimposed backpressure, there is no reason to set a relief device at less than the MAWP. The backpressure at the outlet of every relief device should be such that the device can handle its design capacity with the calculated backpressure under the design relief conditions.

It is common practice to install two relief valves in critical process applications where a shutdown cannot be tolerated. The intent is that if the first relief valve lifts and fails to reseat, a second relief can be switched into service before the first valve is removed for maintenance, without shutting down or jeopardizing the process. This is accomplished by piping the relief valves in parallel and by putting a "car sealed" full port ball or gate block valve on the inlet and outlet of each relief valve. One set of block valves is sealed open and the other sealed closed. ASME-approved selector valves are available, which simplify relief valve switching. This provides an interlock of parallel inlet and outlet block valves and ensures full protection for the process equipment.

Multiple relief valves are required when the relief load exceeds the capacity of the largest available relief valve. It is good practice to install multiple relief valves for varying loads to minimize chattering on small discharges. ASME Sec. VIII, Division 1, 3 and RP 520, Part 1, both stipulate a 10% accumulation above the MAWP for a single relief valve and a 16% accumulation above the MAWP for multiple relief valves. The primary relief valve must be set at or below the MAWP. Supplemental relief valves should have staged pressures. The highest pressure may be set no higher than 105% above the MAWP. If different sized relief valves are used, the smallest relief valve should be set to the lowest pressure.

The most difficult factors for specifying a relief device are determining the limiting cause of pressure relief, determining the relief load and properties of the discharge fluid, and selecting the proper relief device. When the loads are known, the sizing steps are straightforward. RP 520, Part 1, provides formulas for determining the relief valve orifice area for vapor, liquid, and steam relief (API, 2014a). Standard orifices are available by letter designation, orifice area, and body size.

The installation of a relief device requires careful consideration of the inlet piping, pressure sensing lines, and startup procedures. Poor installation may render the relief device inoperable or severely restrict the valve's relieving capacity. Either condition compromises the safety of the facility. Many relief valve installations have block valves before and after the relief valve for in-service testing or removal; however, these block valves must be car sealed or locked open.

Guidelines and consideration are addressed in API RP 520 and the ASME Pressure Vessel Code for the following:
- Inlet piping
- Discharge piping
- Reactive forces
- Tailpipe considerations
- Rapid cycling
- Resonant chatter

B.8.2 Safety Instrumented Systems

A Safety Instrumented System (SIS) detects out-of-control process conditions, and automatically returns the process to a safe state. It is the last line—or near last line—of defense against a chemical process hazard, and it is usually not part of the Basic Process Control System. The last line of defense is what differentiates an SIS from the Basic Process Control System, which is typically used for normal regulatory process control.

As an example, pressure is a potential hazard. Pressure is normally controlled by a regulatory control loop in the Basic Process Control System, which modulates a pressure control valve. An independent high-pressure shutdown function is implemented in a SIS. It is independent in as much as the components used in the BPCS and the SIS are separate, physically and functionally. The SIS is generally independent, both physically and functionally, from the BPCS to ensure that any condition, which might result in an out-of-control process parameter in the BPCS, is safeguarded by the SIS, regardless of the function of the BPCS (Kenexis, 2010).

The SIS will include three types of components including sensor components, a logic solver component, and final control elements and may include a programmable system or a nonprogrammable system.

The SIS presents difficult design challenges. Before the release of risk-based standards for the design of SIS, designs were traditionally implemented using rules of thumb that were quite effective, but not entirely satisfactory. After the implementation of risk-based analysis, users of SIS realized that these engineered safeguards alone had the design flexibility to allow widely diverse designs with widely diverse risk reduction capabilities. The SIS can be designed in a good, better, and best fashion, limiting the amount of risk reduction provided to match the amount of risk reduction needed.

B.8.3 Regulations and Standards

Regulations require compliance with recognized and generally accepted good engineering practices for safety critical controls, including SIS to ensure that industry complies with the norms of operation that have been defined for that industry. Regulations and standards that apply to SIS originated in the late 1980s when industry regulators, including the US Occupational Safety and Health Administration (OSHA) and the US Environmental Protection Agency (EPA), concluded that industry's performance with respect to prevention of major hazards was inadequate. This originated Process Safety Management Standards by OSHA, which were propagated under 29 CFR 1910.119 (OSHA, 2015). In 1996 the EPA propagated its Accidental Release Prevention Program, or 40 CFR Part 68 (US EPA, 2004), which is also called the Risk Management Program. This regulatory standard was substantially similar in the requirements to OSHA's

Process Safety Management Standard. Largely in response to these regulations, the International Society for Automation (ISA) as well as the International Electrotechnical Commission, or IEC, developed industry standards for safety critical control systems to provide additional information on how to comply with OSHA and EPA process safety management regulations.

B.8.3.1 International Electrotechnical Commission 61511

Fig. B.5 presents the SIS Safety Lifecycle as prescribed in IEC 61511 (ISA 84.00.01): "Functional safety—Safety instrumented systems for the process industry sector" (ISA, 1996). The Safety Lifecycle includes a number of specific steps from design through operation, maintenance, testing, and even decommissioning, to address safety throughout the lifetime of a SIS in the petroleum or chemical process.

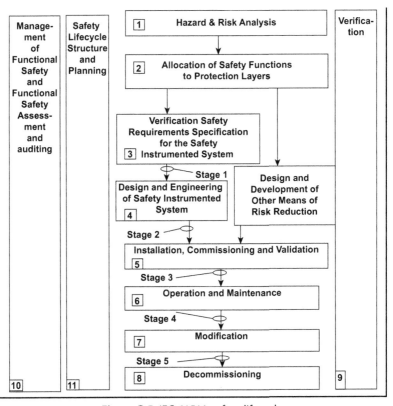

Figure B.5 IEC 61511 safety lifecycle.

The first step in the Safety Lifecycle is "Hazard and Risk Assessment." The premise of this step is that to adequately design a SIS, the hazards against which that system is intended to safeguard must be fully understood.

The next step, "Allocation of Safety Functions," involves assigning certain levels of integrity to each of the safeguards that are used in the process, including Safety Instrumented Functions, as well as non-instrumented functions to achieve an overall level of safety that is acceptable to the company that is operating that process.

The third step in the Safety Lifecycle is "Safety Requirement Specifi-cations." This is an important step to achieve overall functional safety. In the conceptual design of a process, we must make sure that the safety re-quirements are adequately specified before proceeding to other steps in the engineering design lifecycle, including detailed design, construction, installation, and commissioning. This step is where the objectives and means for achieving those objectives are defined. Once completed, the Safety Requirements Specifications (SRS) form the basis for all subsequent design and validation activities.

The steps after SRS (which are often performed in parallel by different groups) are "Detail Design and Engineering" and "Design and Develop-ment" of other non-SIS safeguards. This stage is where the information in the SRS is expanded into more detailed documentation that is used for purchase of equipment, equipment configuration, and installation. This includes tasks such as creating equipment lists, cabinet layout diagrams, internal wiring diagrams, interconnecting wiring diagrams, and PLC programs.

After the detailed design of the SIS is complete, the following step is "Installation, Commissioning, and Validation" of the SIS itself. This stage involves factory acceptance testing, physical installation of the SIS logic solver and all of the field instrumentation, commissioning of those devices, and a validation step that will include site acceptance testing and prestartup acceptance testing.

Once the SIS is installed and operational, a different phase of the life-cycle begins where the design team passes responsibility for the equipment to the operations and maintenance team of the operating company. The "Operation and Maintenance" involves the routine day-to-day interaction with a functioning system. During this phase, operations staff will respond to overt system alarms utilizing procedures written for that purpose. In addition, maintenance will repair the system in response to overt faults and perform periodic function testing to ensure the system's proper operation.

The "Decommissioning" and "Modification" steps are very similar in nature. If changes to the SIS or to the process under control occur, measures must be taken so that the SIS is not compromised in its ability to provide the required amount of risk reduction. During this phase Management of Change (MOC) requirements apply to ensure that when changes are made to the process that are outside the SRS, those changes are adequately analyzed for potential hazards before implementing the change. Decommissioning is a special form of modification where analysis of the removed equipment must be analyzed with respect to its impact on the equipment that will stay in service.

There are three other steps of the Safety Lifecycle that occur over the entire length of an SIS lifetime. These steps are "Management of Functional Safety and Functional Safety Assessment and Auditing," "Safety Lifecycle Structure and Planning," and "Verification." To execute a functional safety project effectively, management of the entire process is important.

Management tasks such as assigning tasks to qualified resources can have a bearing on the project, and thus standard requirements were set. Each step along the way, the standard requires verification that the outputs of each step in the Safety Lifecycle have been achieved, and are consistent with the inputs to that step of the Safety Lifecycle.

B.8.3.2 *International Electrotechnical Commission 61508*
IEC 61511 references IEC 61508 Functional Safety of Electrical/Electronic/Programmable Electronic Safety-related Systems for many items such as manufacturers of hardware and instruments and so IEC 61511 cannot be fully implemented without reference to IEC 61508. IEC 61511 is the process industry implementation of IEC 61508 (IEC, 2010).

B.8.3.3 *Safety Integrity Level*
As per their definition in IEC 61511 (ISA 84.00.01) Safety Integrity Levels (SIL) are order of magnitude bands of average probability of failure on demand (PFDavg). This PFDavg also represents the amount of risk reduction of a preventive safety instrumented function can provide. The ranges for each of the four SIL levels defined in the standard are presented in Table B.6 (IEC, 2004). This figure presents not only SIL in terms of PFDavg, but also Safety Availability and Risk Reduction Factor (RRF). Safety Availability is the complement of PFDavg (i.e., $1 - $ PFDavg), and the RRF is the inverse of PFDavg (i.e., $1/$PFDavg). All of these metrics are commonly used in industry.

Table B.6 Safety integrity level metrics

Safety integrity level	Safety (%)	Probability of failure on demand (%)	Risk reduction factor
SIL 4	>99.99	0.001—0.01	100,000—10,000
SIL 3	99.9—99.99	0.01—0.1	10,000—1000
SIL 2	99—99.9	0.1—1	1000—100
SIL 1	90—99	1—10	100—10

SIL 1 is the lowest level of Safety Integrity that is defined by safety availability by at least 90% up to 99%, essentially providing one order of magnitude of risk reduction. SIL 2 is an order of magnitude safer than SIL 1 in terms of its safety availability. The safety availability of a SIL 1 function would be at least 99% and up to 99.9% safety available. SIL 3 is an order of magnitude on top of SIL 2 in terms of safety availability, and SIL 4 follows accordingly. SIL 4 is rarely seen in the process industries. If a SIL selection process results in a requirement for SIL 4, the user should proceed with care and obtain the assistance of experts.

B.8.3.4 High-Integrity Pressure Protection System

A high-integrity pressure protection system (HIPPS) is a type of SIS designed to prevent overpressurization of a plant. The HIPPS will shut off the source of the high pressure before the design pressure of the system is exceeded, thus preventing loss of containment through rupture of a line or vessel. A HIPPS is considered as a barrier between a high-pressure and a low-pressure section of an installation.

In traditional systems, overpressure is dissipated through relief systems. A relief system will open an alternative outlet for the fluids in the system once a set pressure is exceeded to avoid further accumulation of pressure in the protected system. This alternative outlet generally leads to a flare or venting system to dispose of the excess fluids safely. A relief system aims at removing any excess inflow of fluids for safe disposal, where a HIPPS aims at stopping the inflow of excess fluids and containing them in the system.

Conventional relief systems have disadvantages such as release of process fluids or their combustion products to the environment. With increasing environmental awareness, relief systems are not always an acceptable solution. However, because of their simplicity, relatively low cost, and wide availability, conventional relief systems are still often applied.

HIPPS provides a solution to protect equipment in cases where:
- high pressures or flow rates are processed,
- the environment is to be protected,
- the economic viability of a development needs improvement, and
- the risk profile of the plant must be reduced

HIPPS is an instrumented safety system that is designed and built in accordance with the IEC 61508 and IEC 61511 standards. A HIPPS closes the source of overpressure within 2 s with at least the same reliability as a safety relief valve. Such a HIPPS is a complete functional loop consisting of:
- sensors that detect the high pressure;
- a logic solver, which processes the input from the sensors to an output to the final element; and
- final elements consisting of a valve, actuator, and solenoids that actually perform the corrective action in the field by bringing the process to a safe state by shutting off the source of overpressure.

Fig. B.6 shows three pressure transmitters connected to a logic solver. The solver will decide based on 2-out-of-3 voting whether or not to activate the final element. The 1-out-of-2 solenoid panel decides which valve to close. The final elements consist of two block valves that stop flow to the downstream facilities to prevent exceeding a maximum pressure. The operator of the plant is warned through a pressure alarm that the HIPPS was activated.

This system has a high degree of redundancy:
- Failure of one of the three pressure transmitters will not compromise the HIPPS functionality, as two readings of high pressure are needed for activation.
- Failure of one of the two block valves will not compromise the HIPPS functionality, as the other valve will close on activation of the HIPPS.

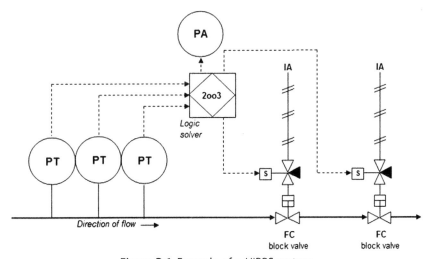

Figure B.6 Example of a HIPPS system.

The European standard EN12186 (formerly the DIN G491) and more specifically the EN14382 (formerly DIN 3381) has been used for the past decades in (mechanically) instrumented overpressure protection systems. These standards prescribe the requirements for the overpressure protection systems, and their components, in gas processing plants. Not only the response time and accuracy of the loop but also safety factors for oversizing of the actuator of the final element are dictated by these standards. Independent design verification and testing to prove compliance to the EN14382 standard is mandatory. Therefore the users often refer to this standard for HIPPS design.

REFERENCES

API RP 520, 2014a. Design and Installation of Pressure Relieving Systems in Refineries, Part I, ninth ed. American Petroleum Institute.
API RP 521, 2014b. Guide for Pressure-Relieving and Depressuring Systems, sixth ed. American Petroleum Institute.
API RP 520, 2015. Design and Installation of Pressure Relieving Systems in Refineries, Part 2, sixth ed. American Petroleum Institute.
ASME, 1971. Fluid Meters — Their Theory and Practice. American Society of Mechanical Engineers.
ASME, 2015. Boiler and Pressure Vessel Code, Sec. 8, Divisions 1 and 2. American Society of Mechanical Engineers.
Beckwith, T.G., 1993. Mechanical Measurements. Addison-Wesley.
Clevett, K., 1986. Process Analyzer Technology. Wiley-Interscience, New York.
Foust, A., et al., 1980. Principles of Unit Operations. Wiley, New York.
IEC 61511-Functional Safety: Safety Instrumented Systems for the Process Industry Sector, 2004. International Electrotechnical Commission.
IEC 61508 — Functional Safety, 2010. International Electrotechnical Commission.
ISA 84.01-Application of Safety Instrumented Systems for the Process Industries, 1996. International Society of Automation.
Jain, R.K., 2011. Mechanical and Industrial Measurements. Khanna Publishers.
Janna, W., 1993. Introduction to Fluid Mechanics. PWS-Kent.
Kenexis, 2010. Safety Instrumented Systems Engineering Handbook. Kenexis Consulting Corporation.
OSHA 29 CFR 1910.119, Occupational Safety and Health Administration, 2015.
Ried, D., 2011. Speed Sensing for Gas & Steam Turbines. Woodward Industrial Turbine Systems.
US Environmental Protection Agency, 2004. 40 CFR Part 68, Subpart G — Risk Management Plan.

INDEX

'*Note:* Page numbers followed by "f" indicate figures and "t" indicate tables.'

Printed in the United States
By Bookmasters